THE ARTHROPODS OF
HUMANS AND DOMESTIC ANIMALS

THE ARTHROPODS OF HUMANS AND DOMESTIC ANIMALS

A GUIDE TO PRELIMINARY IDENTIFICATION

ALAN WALKER
Centre for Tropical Veterinary Medicine
University of Edinburgh
UK

Chapman & Hall
London • Glasgow • Weinheim • New York • Tokyo • Melbourne • Madras

Published by Chapman & Hall, 2–6 Boundary Row, London SE1 8HN, UK

Chapman & Hall, 2–6 Boundary Row, London SE1 8HN, UK

Blackie Academic & Professional, Wester Cleddens Road, Bishopbriggs, Glasgow G64 2NZ, UK

Chapman & Hall GmbH, Pappelallee3, 69469 Weinheim, Germany

Chapman & Hall USA, One Penn Plaza, 41st Floor, New York NY 10119, USA

Chapman & Hall Japan, ITP-Japan, Kyowa Building, 3F, 2-2-1 Hirakawa-cho, Tokyo 102, Japan

Chapman & Hall Australia, Thomas Nelson Australia, 102 Dodds Street, South Melbourne, Victoria 3205, Australia

Chapman & Hall India, R. Seshadri, 32 Second Main Road, CIT East, Madras 600 035, India

First edition 1994
© 1994 Alan Walker

Designed and typeset in Palatino 10/12 by Florencetype Ltd, Kewstoke, Avon
Printed in Great Britain by St Edmundsbury Press, Bury St Edmunds, Suffolk

ISBN 0 412 57280 X

Apart from any fair dealing for the purposes of research or private study, or criticism or review, as permitted under the UK Copyright Designs and Patents Act, 1988, this publication may not be reproduced, stored, or transmitted, in any form or by any means, without the prior permission in writing of the publishers, or in the case of reprographic reproduction only in accordance with the terms of the licences issued by the Copyright Licensing Agency in the UK, or in accordance with the terms of licences issued by the appropriate Reproduction Rights Organization outside the UK. Enquiries concerning reproduction outside the terms stated here should be sent to the publishers at the London address printed on this page.
The publisher makes no representation, express or implied, with regard to the accuracy of the information contained in this book and cannot accept any legal responsibility or liability for any errors or omissions that may be made.

A catalogue record for this book is available from the British Library
Library of Congress Catalog Card Number 94-72646

∞ Printed on permanent acid-free paper, manufactured in accordance with ANSI/NISO Z39.48-1992 and ANSI/NISO Z39.48-1984 (Permanence of Paper)

CONTENTS

Preface	xi
Acknowledgements	xiii
Introduction	xv
How to use this book	xvii

Part One: THE MITES AND TICKS – ACARINA   1
Their biology as an aid to identification   1

1 MITES   3

1.1 Scab mites, mange mites and others – Astigmata   3
    1.1.1 *Psoroptes*   3
    1.1.2 *Chorioptes*   5
    1.1.3 *Otodectes*   6
    1.1.4 *Sarcoptes*   7
    1.1.5 *Notoedres*   8
    1.1.6 *Knemidokoptes*   9
    1.1.7 *Cytodites*   9
    1.1.8 *Laminosioptes*   10
    1.1.9 *Myocoptes*   11
           Stored food mites – *Glycyphagus* and *Acarus*   12
    1.1.10 *Glycyphagus*   12
    1.1.11 *Acarus*   12
    1.1.12 House and bed dust mite – *Dermatophagoides*   13
    1.1.13 Feather mites – Analgidae   14

1.2 Chiggers, follicle mites and others – Prostigmata   15
    1.2.1 Trombiculidae   15
    1.2.2 Follicle mite – *Demodex*   16
    1.2.3 Cat fur mite – *Cheyletiella*   17
    1.2.4 Itch mite – *Psoregates*   17
    1.2.5 *Pyemotes*   18
    1.2.6 *Myobia*   19

	1.3	Bird mites and others – Mesostigmata	20
		1.3.1 *Dermanyssus*	20
		1.3.2 *Ornithonyssus*	21
		1.3.3 *Laelaps*	22
		1.3.4 Halarachnid mites – *Pneumonyssus*	23
	1.4	Beetle mites or Oribatid mites – Cryptostigmata	24

2 TICKS 25

	2.1	Introduction to soft and hard ticks	25
		General features	25
		Life cycles	27
		2.1.1 *Argas* soft ticks	30
		2.1.2 *Ornithodoros* soft ticks	31
		2.1.3 *Otobius* soft ticks	33
		2.1.4 *Boophilus* hard ticks	34
		2.1.5 *Amblyomma* hard ticks	37
		2.1.6 *Hyalomma* hard ticks	39
		2.1.7 *Rhipicephalus* hard ticks	41
		2.1.8 *Dermacentor* hard ticks	43
		2.1.9 *Haemaphysalis* hard ticks	45
		2.1.10 *Ixodes* hard ticks	46
		2.1.11 Other tick genera	48

Part Two: THE INSECTS – INSECTA 51
Their biology as an aid to identification 51

3 FLIES – DIPTERA 53

	3.1	General introduction to the flies	53
	3.2	Nematoceran flies	58
	3.2.1	Phlebotomine sandflies – Psychodidae	58
	3.2.2	Biting midges – Ceratopogonidae	60
	3.2.3	Black flies – Simuliidae	65
	3.2.4	Mosquitoes – Culicidae	68
		General features	68
		Diagnostic features of mosquito genera	70
		Anopheline genera	73
		Anopheles	73
		Culicine genera	73
		Culex	74
		Aedes	74
		Mansonia and *Coquillettidia*	75
		Haemagogus	76

		Sabethes	76
		Psorophora	77
		Culiseta	77
	3.3	Brachyceran flies	78
		3.3.1 Horse flies, deer flies and clegs – Tabanidae	78
		Tabanus	79
		Chrysops	81
		Haematopota	81
		Hybomitra	82
		Lepiselaga	83
		3.3.2 Snipe flies – Rhagionidae	85
		Symphoromyia	85
	3.4	Cyclorrhaphan flies	85
		3.4.1 Introduction to the cyclorrhaphan flies	85
		3.4.2 Louse flies, forest flies and keds – Hippoboscidae	87
		3.4.3 House flies – Musca	90
		3.4.4 Sheep head fly (sweat fly) – Hydrotaea	94
		3.4.5 Sweat flies – Morellia	95
		3.4.6 Eye flies – Hippelates and Siphunculina	95
		3.4.7 Tsetse – Glossina	96
		3.4.8 Stable flies – Stomoxys	99
		3.4.9 Horn flies, buffalo flies and other stomoxyine flies – Haematobia and Haematobosca	101
		3.4.10 Blowflies – Calliphora, Lucilia, Phormia	103
		3.4.11 Screwworm flies – Chrysomya and Cochliomyia	108
		3.4.12 Fleshflies – Sarcophaga and Wohlfahrtia	112
		3.4.13 Tumbu fly and floor maggot – Cordylobia and Auchmeromyia	115
		3.4.14 Nasal bot flies – Oestrus, Rhinoestrus and Cephalopina	119
		3.4.15 Warbles and stomach bots – Hypoderma and Gasterophilus	122
		3.4.16 The torsalo and rodent bots – Dermatobia and Cuterebra	126
4	**LICE – PHTHIRAPTERA**		**131**
4.1	Introduction to the lice		131
4.2	Sucking lice – Anoplura		132
	4.2.1 General features		132
	Human head and body lice – Pediculus		132
	Human pubic louse – Pthirus		133
	Haematopinus		134

		Linognathus	135
		Solenopotes	136
		Polyplax	137
	4.3	Ischnoceran chewing lice	138
		Damalinia (*Bovicola, Trichodectes*)	138
		Trichodectes	139
		Felicola	140
		Goniocotes	140
		Goniodes	141
		Lipeurus	142
		Cuclotogaster	142
	4.4	Amblyceran chewing lice	143
		Menacanthus	144
		Menopon	144
		Heterodoxus	145

5 FLEAS – SIPHONAPTERA — 147

	5.1	Introduction to the fleas	147
		Cat and dog fleas – *Ctenocephalides*	148
		Rat flea – *Xenopsylla*	149
		Human flea – *Pulex*	150
		Sticktight flea – *Echidnophaga*	151
		Rat flea and chicken flea – *Nosopsyllus* and *Ceratophyllus*	152
		Chigger flea – *Tunga*	152

6 BLOOD SUCKING BUGS – HEMIPTERA — 155

	6.1	Introduction		155
		6.1.1	Triatomine bugs	155
			Triatoma	156
			Rhodnius	157
			Panstrongylus	157
		6.1.2	Bed bugs – *Cimex*	159

Part Three: OTHER ARTHROPODS AND METHODS — 161

7 OTHER HARMFUL ARTHROPODS — 163

7.1	Scorpions	163
7.2	Whipscorpions	164
7.3	Spiders	164

		CONTENTS	ix

7.4	Solpugids	165
7.5	Millipedes	165
7.6	Centipedes	165
7.7	Bees	166
7.8	Wasps and hornets	166
7.9	Ants	166
7.10	Beetles	167
7.11	Caterpillars	167
7.12	Moths	168
7.13	Fruit flies	168
7.14	Thrips	169
7.15	Crustaceans	169
7.16	Tongue worms	170
7.17	Delusions of infestation	170

8 METHODS FOR IDENTIFICATION — 171

- 8.1 Collection — 171
  - 8.1.1 Light traps — 171
  - 8.1.2 Suction traps — 172
  - 8.1.3 Animal baited traps — 172
  - 8.1.4 Bait hosts or sentinels — 172
  - 8.1.5 Insect nets and aspirators — 172
  - 8.1.6 Sweep nets and blanket drags — 173
  - 8.1.7 Smell baits and attractants — 173
  - 8.1.8 Visual traps and baffle traps — 174
  - 8.1.9 Electrocutor traps — 174
  - 8.1.10 Sticky traps — 174
  - 8.1.11 Emergence traps — 175
  - 8.1.12 Host searches — 175
- 8.2 Fresh specimens for pathogen isolation, blood-meal tests and age-grading — 176
- 8.3 Preservation — 176
  - 8.3.1 Liquid preservative — 176
  - 8.3.2 Dry preservation — 177
  - 8.3.3 Preservation on microscope slides — 177

x CONTENTS

    8.4   Examination of specimens   178

Appendix A: Summary of clinical signs of direct parasitic damage caused by arthropods   181

Appendix B: Summary of diseases and causative pathogens commonly transmitted by arthropods   183

Appendix C: Glossary   187

Bibliography   201

Index   205

PREFACE

On a global basis the problems caused to humans and their domestic animals by arthropods and the diseases associated with them may be as bad as at any time in history. Although there have been many retreats of medical and veterinary arthropods due to control schemes and general improvement of the living environment of both humans and livestock, there is a constant tendency for the combination of increasing density of human populations and failing economic systems to negate many efforts at control.

The information available on these insects, ticks and mites and the disease pathogens that some of them transmit has been increasing rapidly over the last 20 to 30 years. Much of the most useful parts of this information can be found in the several excellent text books and well illustrated photographic guides that recently have been published on the subject. However, none of this information will be used to good effect unless there are people prepared and willing to use it. The first essential to do something practical in the control of medical and veterinary arthropods is familiarity with the arthropods themselves. Familiarity with what they look like, how they can be distinguished from similar arthropods, and what makes them potentially harmful, is what is needed.

Here lies the problem to which this book is intended to be an answer. A thorough course of practical classes in a well stocked teaching laboratory specializing in all aspects of parasitology is a luxury that few of us who have to deal with these arthropods has ever had. I never had one myself, but for the last 17 years have been helping to deliver two such courses. I have thus learnt the hard way that these arthropods are as confusing as they are fascinating. Because they are so fascinating the confusion over their identification becomes intensely frustrating when even the simplest level of identification has to be made by conventional taxonomic keys written in obscure language and often without illustrations. Formal keys and similar taxonomic information are essential for identification of species. But for the main groups, and usually down to genera, it is possible to use the systems for identification so effectively developed by birdwatchers and other naturalists in their guidebooks written for the general public.

The fundamental element of such systems are the illustrations. The illustrations are purpose drawn. They show at least one view of the whole animal, with the important features for identification fully labelled. Supporting illustrations are given to clarify detailed features. The text then amplifies the illustration with information on other features of importance for identification such as distribution and hosts. The only information given here which is not strictly relevant to identification is that on the disease organisms transmitted by many of these arthropods. However this is not given so that this book can be incorrectly used as a substitute for a full text book on medical and veterinary entomology. It is given as an aid to identification by reference to the mental framework that many users will have who are trained primarily in human and veterinary medicine. Some readers may also expect information on control methods. Again this is the subject of full textbooks, but even some of these do not cover this topic because it is both rapidly changing and of highly localized specialization. Thus it is not given here. However the large list of further reading will help the reader toward this vital topic. The arrangement of the material follows conventional taxonomic systems, supplemented by a Table of classification and detailed list of contents. This should be of use in identification as a mental framework for those with some formal training in systematic biology. The text has been written simply so that it is accessible to as many non-specialist readers as is feasible. However, to not use the routine technical terms is a disservice to the reader's efforts to learn the subject. Thus the large glossary and full index will help guide a way to understanding of technical terms that are needed.

It has been a pleasure to draw and write this book, but my greatest pleasure will be to see it well thumbed on laboratory benches.

Alan R. Walker

ACKNOWLEDGEMENTS

I am greatly indebted to those who have helped and encouraged me to complete this book, particularly my family. Some people acted as anonymous referees, often with substantial help and criticism of various drafts. Others loaned specimens or checked sections of the text and figures. These include: A. Baker, W.N. Beesley, M. Cameron, D.J. Hadrill, M.J.R. Hall, C.H.C. Lyall, R.M. Newson, D.S. Saunders, R.N. Titchener, G.B. White, N. Wyatt. The loan of specimens from The Natural History Museum, London, and the Ashworth Laboratories of the Division of Biological Sciences, University of Edinburgh, is particularly appreciated. A book like this is for the most part derived from the original field and laboratory work of others. It is not appropriate to quote original sources in an identification guide, but the secondary sources are given in the book list. One of the most useful textbooks has been that of Professor D.S. Kettle, to whom I am also greatly indebted for starting me off on my first entomological research in Kenya.

This book was written during research on the biology of ticks at the Centre for Tropical Veterinary Medicine of the Royal (Dick) School of Veterinary Studies, University of Edinburgh; the research being funded by the Overseas Development Administration of the United Kingdom. I am grateful for their support.

INTRODUCTION

This identification guide is intended for use by people who need to identify specimens of insects, ticks, mites and other arthropods that affect the health of humans and domestic animals. These arthropods are particularly interesting because of their parasitic or predatory adaptations. However they are usually studied in order to control them or the diseases associated with them. The key to the long term success of such control is its use in the context of sufficient knowledge of the biology of the arthropods. The starting point for this is correct identification of the arthropods. This is where the problem lies, and it is a great one. Professional taxonomists are few in number and they have to cope with the problem of practical methods for identifying species whilst dealing with one of the fundamental pitfalls of biology. They are confronted with the question – What is a species? Meanwhile those working on the biology and control of the arthropods are often unsure how to identify even the broader groups such as genera and families. This book is intended to encourage people working with arthropods, or the diseases they transmit, to take a wider and more constructive approach to identification.

Identification is likely to be needed for treatment or control of the health problem. This is a job for those in: veterinary and medical diagnostic laboratories; college departments of biology and parasitology; development projects on the control of such pests and of the disease organisms they transmit; research institutes; zoos and wildlife reserves. Specialists in one group of parasitic arthropods may find a guide to other groups useful. Further information on the biology and control of these arthropods is available in textbooks listed in the Bibliography.

No specialist knowledge is required to use this book. It has been written assuming that the reader has some knowledge of biology and may have learnt English as a second language. A glossary of technical terms is given in Appendix C. The use of keys has been avoided because my experience with teaching identification is that they are difficult to use compared to the descriptive approach. However, keys are useful for identification to genus, and essential for species. The use of keys should be the next step for those wishing to extend their competence in identification. The most relevant way to do this is to obtain keys for the local arthropods of concern. The descriptive approach involves much repetition of information, each description being fairly self-contained. Such repetition, or redundancy, is a deliberate aid to easy use of this book.

The level of identification in this book rarely goes beyond genus. Vastly more information is required to identify species and much of this information is incomplete or confusing. It is also easy to confuse harmless arthropods with closely related species which are harmful. Thus identification should be cautious and include evidence of the hosts, distribution and clinical signs. Well preserved specimens (voucher specimens) should be retained for verification. The advice of specialist taxonomists should be sought; preliminary identification will ease communication with specialists.

There is a wide variety of these arthropods so it is necessary to be selective. The main emphasis is on those that transmit the agents of disease and those that are parasitic. These include arthropods that feed from but do not stay on their hosts (sometimes called micropredators), arthropods that live on the outside of their host (ectoparasites) and arthropods that live inside their hosts (endoparasites). Always the relationship is at the expense of the host. Also included are some arthropods causing allergies. Venomous arthropods are listed in Chapter 6. Pests of food, crops and buildings are not included; they tend to have good identification guides produced in the context of large scale pest control operations. Venomous and pest arthropods require different treatment and are best included in books dealing with the wider variety of animals causing problems. Venomous arthropods are well illustrated in *A Colour Atlas of Arthropods in Clinical Medicine* by W. Peters.

Inevitably there will be omissions, distortions and plain errors in this book. I will be pleased to hear from users of this book how it can be improved.

HOW TO USE THIS BOOK

The identifications are based on drawings. With an unknown specimen, run through the book until you find a series of drawings that fit. Check the introductory sections to make sure you are in the right group: mite, tick, fly, louse, etc. Then check in detail the diagnostic features given in the text and on the drawings. Of similar importance is the information on distribution, hosts, behaviour and clinical signs. The information on diseases is included to provide a mental framework for those more familiar with the diseases. This is given as an additional aid to identification. The information on diseases is given cautiously; scientific literature contains many misleading accounts of the isolation of pathogens from arthropods without supporting evidence of their importance as vectors. More information on diseases is available in some of the books listed in the Bibliography.

Many structures will require a stereoscopic (dissecting) microscope or a normal (compound) microscope to see them, but much can be seen with a ×10 magnifying glass. There is a Chapter on Methods for Identification at the end of this book.

The sets of drawings usually contain at least one of the arthropod as an entire specimen and the scale bar shown refers to this. Other features are shown at various convenient scales. The arthropods are shown as they are likely to be presented for examination – usually from a jar of alcohol, or dried, but not as a perfect specimen from a museum. Diagnostic features are emphasized and other features of beauty or interest have been simplified for the sake of clarity. Colours are rarely useful in these identifications and then it is simplest to describe them.

The drawings have all been made from representatives of a single species to illustrate features of the genus. However, the name of the species drawn is rarely given. This is to prevent identification to species without comparative information on similar species.

> *Do not attempt identification to species without the identification keys for the area in which the specimen was found, or without specialist help.*

xviii   HOW TO USE THIS BOOK

Geographical distribution is important for identification but it is rarely useful to list countries. Instead distribution is given by **habitat**, or zoogeographical region and climatic zone, Figures 1 and 2. The climatic zones shown are simplified from those in the *Times Atlas of the World*.

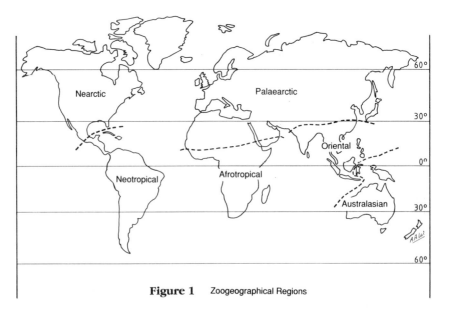

Figure 1   Zoogeographical Regions

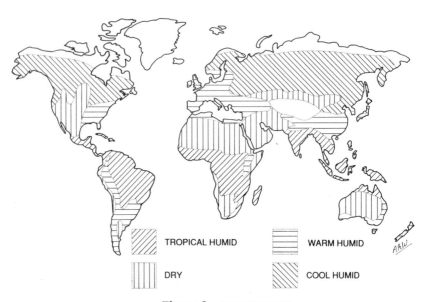

Figure 2   CLIMATIC ZONES

HOW TO USE THIS BOOK xix

- **Tropical humid** climates include rain forest and savanna.
- **Dry** climates include desert and steppe.
- **Warm humid** climates include mediterranean, humid subtropical and marine west coast climates.
- **Cool humid** climates include continental warm summer, continental cool summer and subarctic climates.

The classification, or **taxonomic** grouping, of arthropods of importance to health is complex and uncertain. Table 1 gives a composite version as an outline to help relate the various groups to each other. For a detailed account see *Medical Insects and Arachnids* by Lane and Crosskey.

Table 1 Classification of arthropods of humans and domestic animals

```
Arthropoda (phylum) = mites, ticks, insects, crustaceans, etc.
   Arachnida (class) = mites, ticks, spiders, scorpions
      Acarina (order) = mites and ticks
         Astigmata (or Acaridida) (sub-order) = scab mites etc.
            Sarcoptidae (family) = sarcoptic mites
            Psoroptidae (family) = psoroptic mites
            Cytoditidae (family) = air sac mites
            Analgidae (family) = feather mites
            (plus other families)

         Prostigmata (or Actinedida) (sub-order) = chiggers, etc.
            Trombiculidae (family) = trombiculid mites
            Demodicidae (family) = follicle mites
            Cheyletidae (family) = itch mites
            (plus other families)

         Mesostigmata (sub-order) = bird mites
            Dermanyssidae (family) = bird mites

         Cryptostigmata (sub-order) = oribatid mites

         Ixodida (or Metastigmata) (sub-order) = ticks
            Argasidae (family) = soft ticks
            Ixodidae (family) = hard ticks

      Scorpionida (or Scorpiones) (order) = scorpions
      Pedipalpida (order) = whip-scorpions
      Araneida (or Araneae) (order) = spiders
      Solpugida (order) = solpugids or sun-spiders

   Insecta (class) = flies, lice, fleas, bugs, etc.
      Diptera (order) = two winged flies
         Nematocera (sub-order)
            Culicidae (family) = mosquitoes
            Simuliidae (family) = blackflies
            Psychodidae (family) = sandflies
            Ceratopogonidae (family) = midges
```

            Brachycera (sub-order) = horse flies, clegs
                Tabanidae (family) = horse flies, clegs

            Cyclorrhapha (or the infra-order Muscomorpha (sub-order) = flies
                Hippoboscidae (family) = louse flies
                Muscidae (family) = house flies etc.
                Glossinidae (family) = tsetse
                Calliphoridae (family) = blowflies, etc.
                Oestridae (family) = nasal bot flies, etc
                Cuterebridae (family) = torsalo bots
                Gasterophilidae (family) = stomach bots

        Phthiraptera (order) = lice
            Anoplura (sub-order) = sucking lice
            Ischnocera (sub-order) = chewing lice
            Amblycera (sub-order) = chewing lice

        Siphonaptera (order) = fleas
                Pulicidae (family) = human, cat and rat fleas
                Ceratophyllidae (family) = rodent and chicken fleas
                Tungidae (family) = chigger fleas
                Leptopsyllidae (family) = rodent fleas

        Hemiptera (order) = blood sucking bugs
                Reduviidae (family) = triatomine bugs
                Cimicidae (family) = bed bugs

        Hymenoptera (order) = bees, wasps, ants
        Lepidoptera (order) = butterflies and moths
        Coleoptera (order) = beetles
        Thysanoptera (order) = thrips
Chilopoda (class) = centipedes
Diplopoda (class) = millipedes
Crustacea (class) = crustaceans
Pentastomida (class) = tongue worms

Part One
THE MITES AND TICKS – ACARINA

THEIR BIOLOGY AS AN AID TO IDENTIFICATION

The mites and ticks are within the class Arachnida. This includes the spiders and scorpions, some of which are listed in Chapter 6. Mites and ticks are in the order Acarina (or Acari). There is a very wide variety of mites with most of these being free-living. The taxonomic divisions of mites into sub-orders and families are not generally agreed, but the system used here is a common one. The naming of species of parasitic mites can be a problem; a genus may be split into many species named after the host which each infests, or these may not be considered valid species. Ticks are in the sub-order Ixodida of the order Acarina. They are all parasitic and there are two families, the Ixodidae and the Argasidae.

The outer covering of mites and ticks is called the integument. This consists of a tough and waterproof outer cuticle and a living inner layer, the epidermis. The body is not clearly divided into segments or body parts. The mouthparts are on the body part called the capitulum (or gnathosoma). This resembles a head but if eyes are present they will be on the main body. The mouthparts form a tube. From within this protrude two chelicerae (Figure 1.33g) which are sharp at the tip and cut into the host. The Acarina have no antennae. However the palps on the capitulum have sensory functions, and on the front legs of ticks there are sensory organs (Haller's organs). These legs are used like the antennae of insects as well as for walking.

The rest of the body appears as a single unit like a bag, called the idiosoma. This often has hard plates on the surface and may be greatly expanded when the mite or tick feeds. The larvae of Acarina have three pairs of legs and the nymphs and adults have four pairs of legs. Wings are never present and many mites and ticks remain at the nest or housing of their hosts.

Mites and ticks have a life cycle with an incomplete metamorphosis (Figures 1.1, 1.27, 1.28). The larvae hatch from the egg. After the larvae have fed they develop and moult into nymphs. There may be more than one nymphal stage (or instar). When the nymphs have fed they develop and moult into adults. The adults mate and the females lay one or more batches of eggs. There are many variations to this basic life cycle. In parasitic mites and the ticks, usually all stages are parasitic. However, in the trombiculid mites only the larvae are parasitic and in Otobius ticks only the larvae and nymphs are parasitic. The life cycle is usually rapid in mites (a few days or weeks) and an infestation will usually have all stages present. Tick life cycles often take many months or even several years from one generation to the next.

There are numerous types of free-living mites and a wide variety of parasitic forms. This makes it essential that identification is determined together with clinical information on the infestation. This should include host species, location of infestation on the host and clinical condition of the site of infestation and other related signs.

1 MITES

1.1 SCAB MITES, MANGE MITES AND OTHERS – ASTIGMATA

This group is called the Astigmata because they have no stigmata or air breathing pores. Another name is the Acaridida. They are all very small. Most species are free-living, some are pests of stored products and some are specialized parasites. Most of the parasitic mites are very similar. However it is very useful for identification to know what host and where on that host the mites came from.

Other important features for identification are the suckers on the legs and the spines on the body. The sucker is similar to the pulvillus of hard ticks. It is often called the pretarsus, composed of the stalk or pedicel and the adhesive caruncle at the end. The first segment of the leg (coxa) often has an inner extension along the body wall for muscle attachment. It is called the apodeme and is an important feature for identification. Most of the astigmatid mites described here are parasitic. In addition there are some mites that are not parasitic but are of medical importance because they produce allergic effects. Examples of these are the stored food mites and house dust or bed dust mites. There is a very wide variety of species of this type of mite.

1.1.1 *Psoroptes*

Structure

- The body is oval and most legs are long and protrude beyond the edge of the body.
- The integument is striated.
- The third legs of females have long thick setae extending from the last segment.
- Mouthparts are pointed and adapted for piercing the skin.

4 MITES

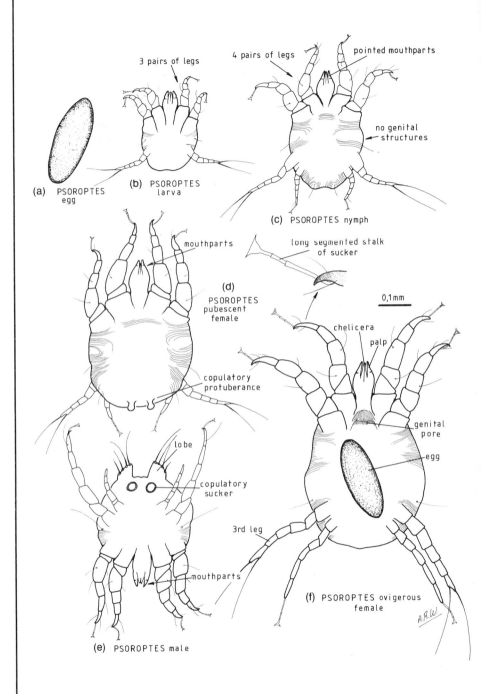

Figure 1.1

SCAB MITES, MANGE MITES AND OTHERS – ASTIGMATA

- There are no spines on the dorsal surface.
- The male has a pair of lobes of curved outline at the back of the body, and a pair of copulatory suckers.
- Females are in two forms: the pubescent female has copulatory protuberances and when this form has mated it moults to the ovigerous female which develops the eggs.
- Leg suckers have long and segmented stalks (pedicels).
- Females have suckers on legs 1, 2 and 4.
- Males have suckers on legs 1, 2 and 3.

Other features and disease

Psoroptes mites feed at the external surface of the host skin, causing inflammation, exudation and scab formation. They are known as scab mites. *Psoroptes ovis* infests sheep, with infestations spreading from the shoulders and back line. Heavy infestations cause severe stress and loss of wool. The same species also may infest cattle and horses. *Psoroptes cuniculi* infests the ears of rabbits causing canker or otitis, and will also infest ears and other areas on sheep and goats, horses and donkeys. There are species named after deer, horses and donkeys. *Psoroptes* mites are found worldwide.

1.1.2 *Chorioptes*

Structure

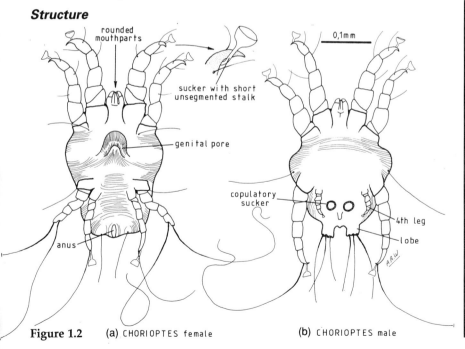

Figure 1.2 (a) CHORIOPTES female (b) CHORIOPTES male

6 MITES

- The main body features are similar to *Psoroptes*.
- The integument is striated.
- The mouthparts are rounded in outline and are adapted for chewing.
- Males have a pair of lobes of square ended shape at the back of the body, and a pair of copulatory suckers.
- The fourth legs of females are long, those of the male are short.
- Leg suckers have a short stalk without segments.
- Females have suckers on legs 1, 2 and 4.
- Males have suckers on legs 1, 2, 3 and 4, but suckers of legs 4 are very small.

Other features and disease
Chorioptes bovis feeds at the external surface of the skin causing inflammation, exudation and scab formation. This species infests cattle, starting from the base of the tail or the legs and tending to spread to the back and neck. It also infests sheep, goats and horses. Heavy infestations cause scabbing of the skin and loss of condition which can lead to emaciation and damage to hides.

1.1.3 *Otodectes*

Structure

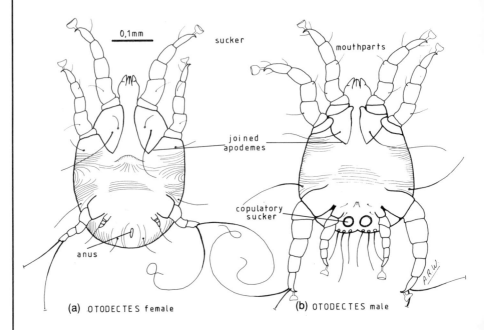

(a) OTODECTES female   (b) OTODECTES male

Figure 1.3

SCAB MITES, MANGE MITES AND OTHERS – ASTIGMATA

- The main body features are similar to *Psoroptes* and *Chorioptes*.
- The integument is striated.
- Males have copulatory suckers but the end of the body does not have distinct angular lobes.
- The apodemes extending from coxae of legs 1 and 2 are joined.
- The fourth legs of females are very small and without suckers.
- Leg suckers have short stalks without segments.
- Females have suckers on legs 1 and 2.
- Males have suckers on legs 1, 2, 3 and 4.

Other features and disease
Otodectes cynotis feeds at the external surface of the skin causing inflammation and exudation. Infestations are usually restricted to the ears of cats and dogs, causing parasitic otitis.

1.1.4 *Sarcoptes*
Structure

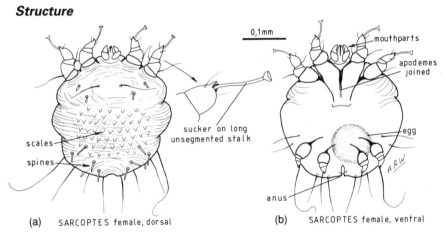

(a) SARCOPTES female, dorsal   (b) SARCOPTES female, ventral

Figure 1.4

- The body outline is circular; the legs are short with only legs 1 and 2 protruding beyond the outline of the body.
- The dorsal surface has angular scales and thick spines.
- The integument is striated.
- The apodemes from the coxae of the first legs are joined centrally in a Y shape.
- The anus is at the posterior end of the body.
- Leg suckers are long and without segments.
- Females have suckers on legs 1 and 2.

8 MITES

- Males have suckers on legs 1, 2 and 4.

Other features and disease

Sarcoptes scabiei burrows into the epidermis of its host forming long tunnels in which the life cycle is completed. This causes intense irritation, inflammation, thickening of the skin and scab formation. Secondary immune reactions may produce a rash at sites away from the infested area. Humans are easily infested, particularly in crowded and poor living conditions. Infestation of pigs is common, producing sarcoptic mange, leading to reduced productivity. Sarcoptic mange is often a severe problem in camels and can be fatal. Infestation also occurs on sheep, goats, equines, and dogs. Infestation may start in a wide variety of sites, often the head but also the groin and back.

1.1.5 *Notoedres*

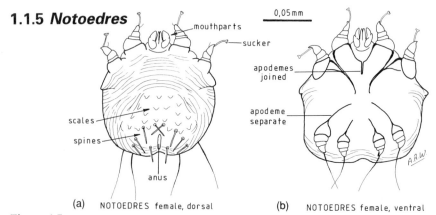

Figure 1.5

Structure

- Main body features are similar to *Sarcoptes*, but the dorsal surface has less distinctly angular scales and the anus is dorsal.
- The integument is striated.
- The apodemes from the coxae of the first legs are joined centrally in a Y shape.
- Leg suckers are long and without segments.
- Females have suckers on legs 1 and 2.
- Males have suckers on legs 1, 2 and 4.

Other features and disease

Notoedres cati burrows in the epidermis of cats forming mange. It also infests dogs and rabbits. Infestations start on the head.

SCAB MITES, MANGE MITES AND OTHERS – ASTIGMATA

1.1.6 *Knemidokoptes*

Structure

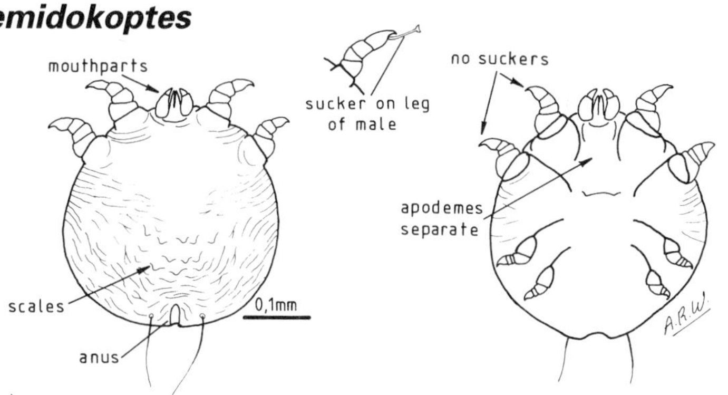

Figure 1.6 (a) KNEMIDOKOPTES female, dorsal (b) KNEMIDOKOPTES female, ventral

- This genus is also spelt *Cnemidocoptes* or *Knemidocoptes*.
- Main body features are similar to *Sarcoptes*, but the dorsal surface has only faint and irregular scales, no spines and the anus is dorsal.
- The integument is striated.
- The apodemes from the coxae of the first legs of females are separate; in males they are joined.
- Leg suckers occur only on the male, on all legs, and are long and unsegmented.

Other features and disease

Knemidokoptes mutans and other species burrow into the epidermis of the legs of birds. In poultry and other domestic birds the scales of the legs are lifted up by the inflammation and swelling of the leg. The resulting deformity and scabbing is known as scaly leg. Secondary complications may kill the birds. *Knemidokoptes gallinae* infests the skin near the base of feathers, causing great irritation and leading to depluming. *Knemidokoptes pilae* causes scaly face in pet cage birds.

1.1.7 *Cytodites*

Structure

- The integument is smooth without spines or striations and with few setae.
- The mouthparts are reduced to a tube.
- Legs are well developed and protrude beyond the body.
- The apodemes from the coxae of the first legs are joined centrally in a Y shape.
- Legs of the female end in pretarsi with a small sucker.
- Legs of the male have no suckers.

10 MITES

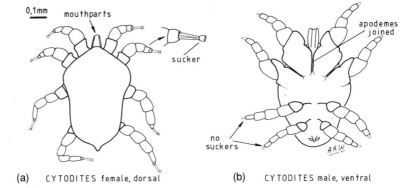

(a) CYTODITES female, dorsal (b) CYTODITES male, ventral
Figure 1.7

Other features and disease
Cytodites nudus, the air-sac mite, is a specialized parasite of the lungs and airsacs of birds. It infests poultry in North America and South Africa and may cause loss of production.

1.1.8 *Laminosioptes*

Structure

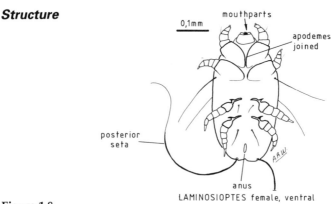

LAMINOSIOPTES female, ventral
Figure 1.8

- The body is elongate with well developed legs which protrude beyond the body.
- The integument is mainly smooth, with few setae but there are two long setae from the posterior end of the body.
- The mouthparts are short.
- The apodemes from the coxae of the first legs are joined centrally in a Y shape.
- The legs end in pretarsi but these do not have obvious suckers.

Other features and disease

Laminosioptes cysticola, the fowl cyst mite, infests the subcutaneous tissue of poultry. Nodules form, reducing the value of the carcass. This species occurs worldwide. A genus in a related group is *Pneumocoptes*; these infest the lungs of rodents.

1.1.9 *Myocoptes*

Structure

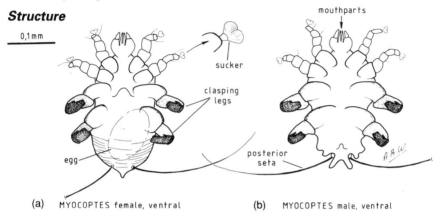

(a) MYOCOPTES female, ventral (b) MYOCOPTES male, ventral

Figure 1.9

- This mite has highly characteristic third and fourth legs adapted for clasping to the hairs of its host.
- The integument is striated.
- The mouthparts are well developed.
- Legs 1 and 2 end in short suckers.
- The posterior end of the body has two long setae.

Other features and disease

Myocoptes musculinus infests the fur of wild and laboratory mice.

STORED FOOD MITES – *GLYCYPHAGUS* AND *ACARUS*

1.1.10 *Glycyphagus*

Structure

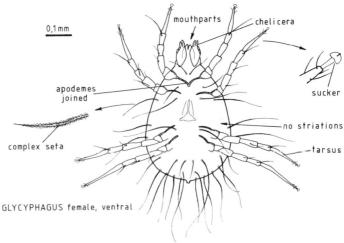

Figure 1.10

- The chelicerae of the mouthparts are well developed.
- The body is oval, the body surface is rough with many minute spines but without striations.
- The legs are long, particularly the last segment (tarsus).
- The legs end in suckers.
- The apodemes from the coxae of the first legs are joined centrally in a Y shape.
- There are numerous long setae and these have many finer setae on them.

1.1.11 *Acarus*

Structure

- The chelicerae of the mouthparts are well developed.
- The body is oval in outline.
- The integument is rough, but without striations.
- The legs are long and end in a claw.
- The apodemes from the coxae of the first legs are joined centrally in a Y shape.
- *Acarus siro* has few short setae.
- The related *Tyrophagus putrescentiae* is similar to *Acarus* but has longer setae.

SCAB MITES, MANGE MITES AND OTHERS – ASTIGMATA

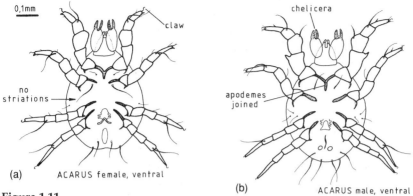

Figure 1.11

Other features and disease of grain mites

Glycyphagus domesticus, *G. destructor*, *Acarus siro*, *Tyrophagus putrescentiae* and other related species are found in stored foods, grain and hay. These mites are not parasitic although they may bite humans. They feed on, or in association with, the stored foods on which they are found. Workers handling these foods may develop allergic reactions to the mites and suffer from pruritus, dermatitis, rhinitis or asthma. These symptoms have various names such as grocer's itch (caused by *Tyrophagus*) and baker's itch (caused by *Acarus*).

1.1.12 House and bed dust mite – *Dermatophagoides*

Structure

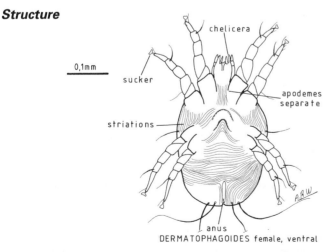

Figure 1.12

- The mouthparts are well developed, with distinct chelicerae.
- The body is oval with long legs and the legs end in suckers.
- The apodemes from the coxae of the first legs are separate.
- The integument is striated.
- There are few setae, but there are four long ones from the posterior body margin.

Other features and disease
Dermatophagoides pteronyssinus is found in human housing, particularly in bedding and furniture, in many parts of the world. It feeds on epidermal flakes shed from human skin. It is not parasitic but this species, and also *D. farinae* and *Euroglyphus maynei*, produce material to which some humans become allergic, causing rhinitis or asthma.

1.1.13 Feather mites – Analgidae (*Megninia*)
Structure

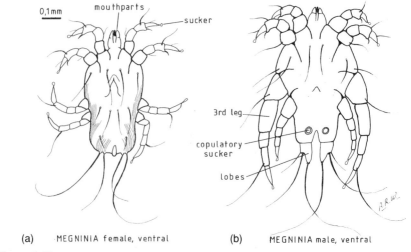

(a)　　MEGNINIA female, ventral　　(b)　　MEGNINIA male, ventral

Figure 1.13

- These are astigmatid mites found on the feathers of birds.
- The legs end in suckers with short stalks.
- The female has all legs of similar size.
- The male has greatly enlarged third legs and large posterior lobes with copulatory suckers.

Other features and disease
Infestations by *Megninia* species may lead to depluming in poultry.

1.2 CHIGGERS, FOLLICLE MITES AND OTHERS – PROSTIGMATA

This group is called the Prostigmata because the stigmata or air breathing pores are at the front of the body, on or near the mouthparts. Another name is the Actenidina. The stigmata are difficult to see. There are many varieties of free-living types, and a few varied parasitic types.

1.2.1 Trombiculidae

Mites of this family are called chigger mites, scrub itch mites, harvest mites, akamushi, etc. They should not be confused with chigger fleas (*Tunga*). Only the larvae are parasitic. The adults are free-living mites which look like very small **hairy** spiders.

Structure

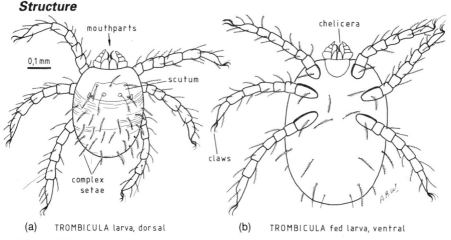

(a) TROMBICULA larva, dorsal (b) TROMBICULA fed larva, ventral

Figure 1.14

- Larvae have three pairs of legs and these end in claws.
- Larvae have large mouthparts with conspicuous chelicerae.
- The larva has a red or orange colour.
- The dorsal surface has a scutum.
- Both surfaces of the body have complex setae which have smaller setae on them.

Other features and disease

There are more than 3000 species of trombiculids, in many genera. Some important genera are: *Trombicula, Leptotrombidium, Euschoengastia, Neoschoengastia*. The distribution is worldwide. The larva occur on vegetation and attach to a wide variety of mammals and other vertebrate hosts. They feed by forming a feeding tube or stylostome, in the skin of the host. They detach after a few days but the reaction in the skin may cause intense irritation and dermatitis in humans and livestock. The larvae may form clusters on the host, causing damage to the skin, for example *Neoschoengastia americana* which feeds on turkeys in North America.

Species of *Leptotrombidium* transmit *Rickettsia tsutsugamushi* causing scrub typhus in humans, often known as tsutsugamushi disease or chigger-borne rickettsiosis. This disease occurs in the Oriental Region.

1.2.2 Follicle mite – *Demodex*

Structure

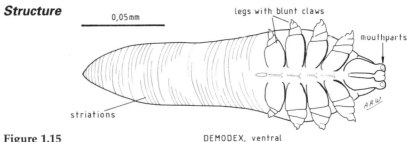

Figure 1.15 DEMODEX, ventral

- These are small mites specialized for living in the **hair follicles** and sebaceous glands of many types of mammals. Their appearance is highly characteristic.
- Adults have four pairs of very short legs ending in small blunt claws.
- The main body is elongated, with fine striations around the posterior part.
- There are few or no setae.

Other features and disease
The life cycle is entirely within the skin of the host. Infestation with *Demodex* (**demodecosis**) is very common. The species *D. follicorum* and *D. brevis* infest the face of most humans, but without being noticed. There is usually no irritation or pathological condition but in some individuals the numbers and spread of the mites on the hosts increase to form clinical demodecosis. On cattle this occurs as flat nodules in the skin with a massive enlargement of the sebaceous glands which contain vast numbers of *D. bovis* mites. In dogs, *D. canis* infestation may become generalized as a squamous form, with loss of hair and thickening of the skin. A more severe form in dogs is known as pustular and is complicated by secondary bacterial infection. However, demodecosis does not cause pruritus. Other species of *Demodex* infest all domestic and many other mammals.

CHIGGERS, FOLLICLE MITES AND OTHERS – PROSTIGMATA

1.2.3 Cat fur mite – *Cheyletiella*
Structure

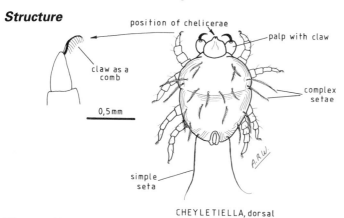

Figure 1.16

- Adults have four pairs of legs, these end in claws with a structure like a comb.
- Dorsal and ventral surfaces have complex setae which have smaller setae on them.
- The body is slightly constricted to form a waist.
- There are two long simple setae projecting from the posterior end of the body.
- Mouthparts are large, with palps which end in large inwardly curved claws.

Other features and disease
These mites live on the outer layers of the epidermis of their hosts, completing their whole life cycle on the host. Several species infest cats, dogs and rabbits. The mites walk about actively and are liable briefly to infest humans, who react to these mites with irritating spots.

1.2.4 Itch mite – *Psorergates*
Structure
- These are small mites with the legs arranged in a circular pattern around the body.
- Legs are short, stout, and end in small claws.
- The outlines of the first coxae form a pair of hook shapes.
- Some segments of all legs protrude as long spine shapes.
- At the posterior margin of the body the female has two pairs of long setae, and the male has one pair.

18 MITES

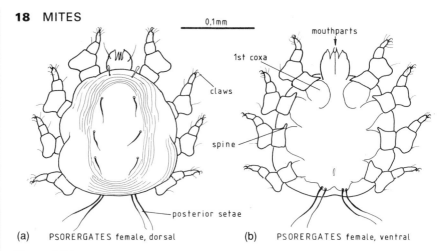

Figure 1.17

Other features and disease
Psorergates ovis (also known as *Psorobia ovis*) infests sheep causing irritation. Grooming by the sheep leads to damage to the wool. The infestations are superficial, the mites live on the skin surface. *Psorergates bos* infests cattle and there are other species on other mammals. *Psorergates* occurs on domestic animals in Australasian, Afrotropical and Nearctic Regions.

1.2.5 *Pyemotes*
Structure

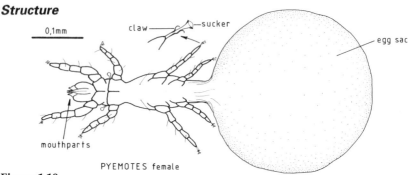

Figure 1.18

- These are small mites with an elongated body and small mouthparts.
- The front two pairs of legs are separated from the back two pairs by a long space.
- The legs end in a pair of small claws and a sucker.
- The posterior end of the female body forms into a large egg sac when eggs are being produced.

Other features and disease
Several species of *Pyemotes* naturally infest beetles and other insects associated with stored food products or furniture. *Pyometes tritici* is known as the straw (or grain) itch mite; it causes dermatitis in persons handling these products.

1.2.6 Myobia

Structure

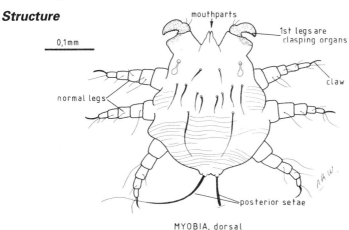

MYOBIA, dorsal

Figure 1.19

- These are small rounded mites with reduced mouthparts.
- The integument is striated.
- The first pair of legs are formed into conspicuous clasping organs.
- The other legs are normal and end in a single large claw.
- There is a pair of long thick setae at the posterior margin of the body.

Other features and disease
Mites of the family Myobiidae infest rodents and other mammals. They may cause problems in laboratory colonies of rodents.

20 MITES

1.3 BIRD MITES AND OTHERS – MESOSTIGMATA

This group is called Mesostigmata because the stigmata (air breathing pores) open at the middle of the body. Another name is Gamasida, or gamasid mites. They are also called the peritreme mites because from the stigmata, long grooves or tubes called peritremes project forward (Figure 1.21b). Peritremes are visible beneath the legs in mites which have been prepared on a microscope slide so that they are semi-transparent. Many of the species are large and free-living. However there are some highly specialized parasites which infest the respiratory system of mammals and birds: *Pneumonyssoides caninum* of dogs; *Pneumonyssus simicola* of monkeys; *Sternostoma* species in birds. Less specialized is *Raillietia auris* which infests the outer ear canal of cattle.

1.3.1 *Dermanyssus*

Structure

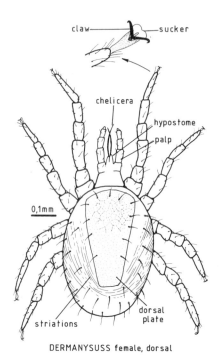

DERMANYSUSS female, dorsal

Figure 1.20

- Legs are long and at the end are pairs of claws and a pad-like sucker (caruncle).
- The hypostome of the mouthparts projects forward conspicuously. From it two long and sharp chelicerae may be visible if extended.

BIRD MITES AND OTHERS – MESOSTIGMATA

- The palps are large and have five segments.
- There is a dorsal plate with a rectangular shape, its posterior end is angular and blunt.
- The ventral surface has a large rounded genito-ventral plate and an anal plate of rounded triangular shape.
- The stigmata are between coxae of legs 3 and 4 and the peritremes are long.
- The integument is striated and the plates have complex patterns of angular shapes on their surface.
- There are distinct spines on the dorsal surface.

Other features and disease

Dermanyssus gallinae, the red chicken mite, infests the nests of poultry and other birds from which it goes onto the birds to feed. Infestations can be of very large numbers of mites causing stress and loss of production. *Dermanyssus gallinae* and other species will leave old nests without birds and crawl long distances searching for hosts which may include humans. A related species is the house mouse mite, *Allodermanyssus* (or *Liponyssoides*) *sanguineus*, which may feed on humans when rodents have been cleared from houses, it transmits *Rickettsia akari* causing rickettsialpox in humans.

The cattle ear mite, *Raillietia auris*, is similar in structure to other free-living mesostigmatid mites; it is found in the outer ear canal of cattle usually causing no harm.

1.3.2 *Ornithonyssus*

Structure

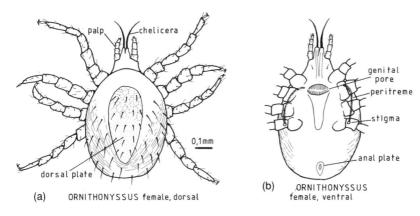

(a) ORNITHONYSSUS female, dorsal
(b) ORNITHONYSSUS female, ventral

Figure 1.21

22 MITES

- These mites may be known as *Liponyssus*.
- They are similar to *Dermanyssus* but are usually more spiny.
- The dorsal plate is oval shaped with a pointed posterior end.
- The anal plate has a shape similar to the dorsal plate.
- The integument is striated and the plates have complex patterns of angular shapes on their surface.

Other features and disease

Ornithonyssus bursa, the tropical fowl mite, is a very common parasite of poultry and other birds in warm humid and tropical humid climates of all tropical regions (pantropical). Similarly *O. sylviarum*, the northern fowl mite, feeds on poultry in cool humid climates, both north and south of the tropics. These mites live on their hosts and do not survive for long off the host. The tropical rat mite, *O. bacoti*, is common on rats and may infest laboratory rats. These mites live in the nests of their hosts, and when nests are deserted they will crawl to find other hosts. Humans are bitten and *O. bacoti* is possibly involved in the transmission of *Rickettsia akari*, causing rickettsialpox.

1.3.3 *Laelaps*

Structure

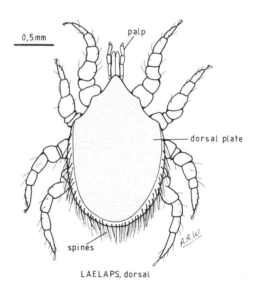

Figure 1.22

- These mites are basically similar to *Dermanyssus*.
- The mouthparts protrude forward conspicuously and the palps have five segments.

- The dorsal plate covers most of the dorsal surface.
- Legs are stout.
- There are numerous thick setae or bristles protruding from the sides of the body.

Other features and disease
This genus is typical of numerous mites which normally infest rodents but which may cause biting nuisance to humans. Other similar genera are: *Haemolaelaps*, *Haemogamasus*, *Androlaelaps*. These mites have a worldwide distribution.

1.3.4 Halarachnid mites – *Pneumonyssus*
Structure

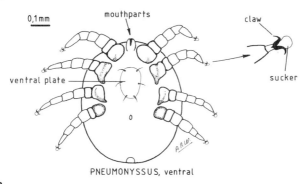

Figure 1.23

- The family Halarachnidae are specialized for **endoparasitic** life in their hosts.
- Mouthparts are small.
- There are few spines or setae.
- The legs end in claws and suckers.
- Dorsal and ventral plates are present but not conspicuous.
- *Pneumonyssus simicola* mites are found in the lungs of monkeys and baboons.
- *Pneumonyssoides caninum* mites are similar to *P. simicola* but have larger mouthparts and are found in the nasal sinuses of dogs.

Other features and disease
Infestations with *Pneumonyssus* or *Pneumonyssoides* do not usually cause clinical disease, but under conditions of stress infestations may increase to cause serious inflammation of the nasal sinuses, bronchi and lungs. Death may result in severe cases.

1.4 BEETLE MITES OR ORIBATID MITES – CRYPTOSTIGMATA

This group is known as the Cryptostigmata because they have concealed stigmata and are covered in large plates giving them the appearance of minute beetles. Another name is the Oribatida.

Structure

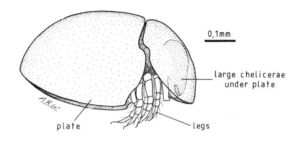

PSEUDOTRITIA, lateral

Figure 1.24

- Mouthparts may be visible and are well developed with chelicerae shaped like pincers and with large palps.
- Legs end in one or more large claws but there are no suckers.
- A typical genus, *Pseudotritia*, is shown in Figure 1.24.

There are many other similar genera of cryptostigmatid mites, the information given here is not sufficient to identify a mite as *Pseudotritia* or other genus.

Other features and disease

Crypostigmata are free-living non-parasitic mites found in leaf litter and soil. They can be the intermediate hosts of *Moniezia*, a tapeworm parasitic in cattle, sheep and goats.

2 TICKS – IXODIDA

2.1 INTRODUCTION TO SOFT AND HARD TICKS

General features

Ticks are usually mentioned separately from mites but they are another sub-order of the Acarina, known as the Ixodida. An uncommon name for the sub-order is Metastigmata, because they have large stigmata in the form of spiracles. Ticks are all parasitic on vertebrate animals and can feed only on such hosts. They spend long periods away from their hosts, amongst vegetation or in cracks in soil, crevices or buildings. Thus they are highly adapted for survival without food and many are very resistant to desiccation. The outer layer of ticks, known as the integument, is waterproofed with wax.

Ticks find hosts by waiting on vegetation, or by crawling rapidly when they sense them with the Haller's organs in the front legs. Ticks feed through the skin of their host so they are usually visible, but some species occur deep in the external ear. The larvae and nymphs in the life cycle are small and often difficult to detect. Reactions to ticks may include generalized toxic or paralytic effects and abscessation at the feeding site. They do not cause generalized dermatitis such as the mange or scab produced as a reaction to mites.

Ticks are divided into two families, the Ixodidae or hard ticks and the Argasidae or soft ticks (Figures 2.1 and 2.2). The hard ticks have a hard shiny scutum like a shield on the dorsal surface and in the males this covers most of the dorsal surface of the tick. The soft ticks do not have this hard scutum, but their integument is not soft in the usual sense, it is very tough and of rough texture. The mouthparts of ticks consist of a central hypostome which is the piercing and sucking tube, and a pair of palps which do not pierce the skin of the host. Other differences between the hard ticks and soft ticks are shown in Figures 2.1 and 2.2. Hard ticks have the mouthparts visible in front of the body, all stages have adhesive pulvilli on the legs, the spiracle is large and behind the fourth legs. Soft ticks have the mouthparts ventral (except some larvae), the nymphs and

26 TICKS

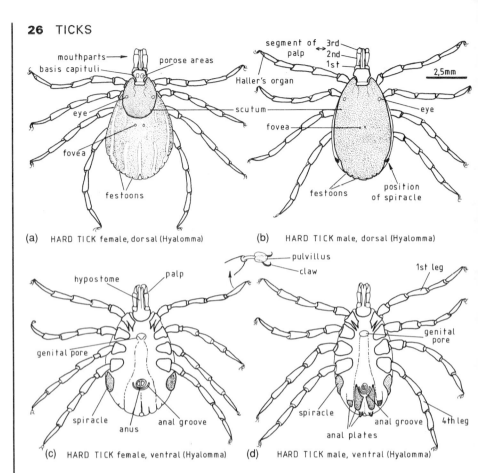

Figure 2.1

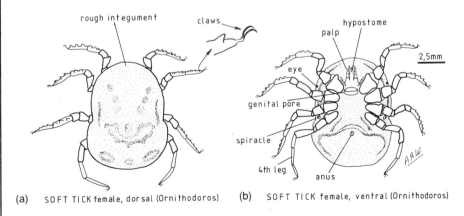

Figure 2.2

adults have no pulvilli, the spiracle is small and in front of the fourth legs. The sexes in soft ticks are similar except for the **genital pore**. Some hard ticks, particularly *Amblyomma* and *Dermacentor* (Figures 2.10b, 2.13a) have patterns of light coloured pigment on the scutum. It is important not to mistake the dark colour of the gut which may show through the scutum (Figure 2.9c) for these patterns.

Life cycles

The life cycles of hard ticks are varied but of two basic types (Figures 2.3 and 2.4). Most hard ticks will feed on three individual hosts separated by long periods in the vegetation whilst moulting to the next stage or laying eggs (Figure 2.3). Whilst on the host any one stage may look like a flat unfed tick or be expanded by about five to ten times its original size. This depends on how much blood and lymph it has drunk. Feeding takes a few days to several weeks. When it detaches, the tick's integument forms a case inside which the next stage develops. Female ticks take an enormous meal of blood and mate once. They drop to the ground, then slowly lay thousands of eggs in one batch, after which they die. Males usually remain on the host for more matings.

Some ticks have modified this cycle so that the larvae and nymphs stay on one host and the adults feed on a second host (two-host cycle). Other ticks have all stages feeding one after the other on one individual animal (one-host cycle) so the only stages off the host are the females and the larvae (Figure 2.4).

Soft ticks feed in minutes rather than days. There may be several nymphal stages, and the females and males take repeated small-blood meals with the female laying small batches of eggs after each blood-meal (Figure 2.5). This is typical of the genus *Ornithodoros* but the other types of soft ticks have different life cycles, details of which are given later.

Most species of soft tick and a few species of hard tick have a life cycle centred on the nest or buildings in which the hosts live. Thus the ticks may be easier to find there than on the hosts.

Larvae and nymphs are more difficult to identify to genus than adults because the features are smaller or less distinct. However it is worth attempting to identify immature ticks:

- Larvae of hard ticks are obvious with only three pairs of legs, a scutum at the anterior end and no genital pore or spiracle.
- Nymphs of hard ticks have four pairs of legs, a scutum at the anterior end and no genital pore, but they do have spiracles.
- Larvae of soft ticks have only three pairs of legs, the mouthparts are visible from in front and small pulvilli may be visible on the legs.
- Nymphs of soft ticks resemble the adults but without the genital pore.

28 TICKS

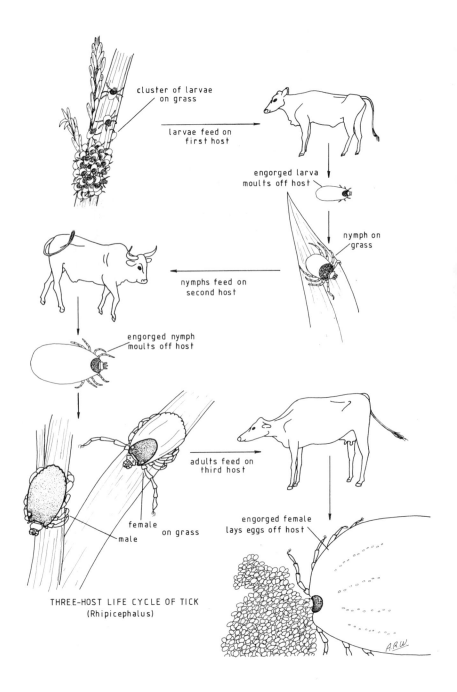

Figure 2.3

INTRODUCTION TO SOFT AND HARD TICKS 29

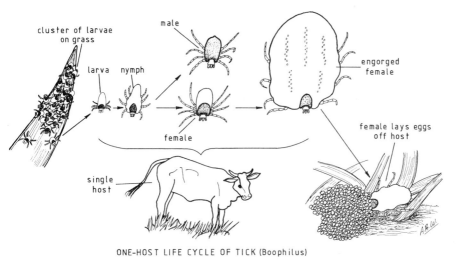

Figure 2.4

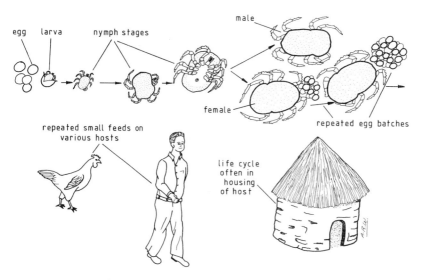

Figure 2.5

2.1.1 *Argas* soft ticks

This is a genus of about 50 species, known as fowl ticks or tampans.

Structure

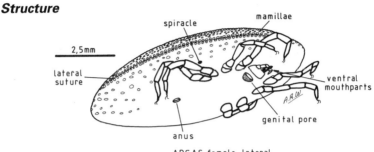

Figure 2.6

- The nymphs and adults have typical soft tick features.
- The larvae have small pulvilli and their mouthparts project forward.
- From above the body outline is oval or circular, from the side they have a shape like two dinner plates stuck together forming a ridge (lateral suture) round the middle.
- The integument has patterns of flat areas mixed with rounded bumps (mamillae) and fine complex folds.
- Mouthparts are well developed for blood sucking in all stages.

Sexes
These are similar except for minor differences in the shape of the genital pore.

Distribution
Argas species are widely distributed in all climatic zones and zoogeographical Regions, strongly preferring dry habitats. They are a common pest of poultry in tropical countries but are more commonly found in wild bird nests and bat caves in warm and cool humid climates throughout the world. *Argas persicus* has spread into warm and cool humid climates on poultry. All stages are found off the host in soil, sand, cracks or nests.

Hosts
Typical hosts are birds and bats. *Argas persicus*, the fowl tick, possibly bites humans. The pigeon tick *Argas reflexus* bites humans in contact with old nests or bird housing. Other *Argas* species feed on mammals and reptiles but are not pests of livestock.

Life cycle

Larvae seek hosts and attach for a few days to weeks to take one large blood-meal, then they detach and moult to nymphs. There are two nymphal stages and a final moult to the adults. Nymphs and adults take repeated small blood-meals lasting a few minutes to hours and the females lay about five small batches of eggs. The life cycle is completed in four to five weeks but if no hosts are available adult ticks survive for years without feeding.

Behaviour

Argas ticks will crawl from their resting places at night in search of hosts. During the day they rest in cracks in buildings or in soil. They have a strong tendency to squeeze into narrow spaces.

Disease

HUMANS Toxic symptoms may occur if these ticks (for example *Argas reflexus*) feed on humans.

DOMESTIC ANIMALS Large infestations cause loss of production or death in poultry. The rickettsial organism *Aegyptianella pullorum* is transmitted to poultry. The bacterium *Borrelia anserina*, causing avian spirochetosis, is transmitted by *Argas persicus*.

2.1.2 *Ornithodoros* soft ticks

This is a genus of about 90 species.

Structure

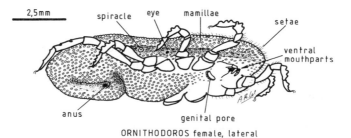

Figure 2.7

- The nymphs and adults have typical soft tick features.
- From above the outline is oval and unless the tick is expanded with a recent blood-meal there are large deep folds in the integument.
- There is no lateral suture as in *Argas*.

- The integument has numerous small bumps (mamillae) and fine setae but these are distinct from the spines of *Otobius*.
- In some species there are small eyes in a lateral position.

Sexes
These are similar except for small differences in the genital pore.

Distribution
Ornithodoros are found in all zoogeographical regions. They prefer dry habitats; they may be found in areas with high rainfall but select drier sites in human and animal housing, particulary mud huts. They commonly infest animal burrows, and prefer to hide in narrow cracks or under sand.

Hosts
These are a wide range of mammals, including man; livestock most affected are pigs, camels and cattle. Poultry may be bitten, particularly by nymphs of the *Ornithodoros moubata* complex of species.

Life cycle
This is similar to *Argas*. The larvae either feed rapidly, or in *Ornithodoros moubata* and *O. savignyi*, the larvae remain within the open egg shell without feeding until they moult to nymphs. The life cycle can be as short as two months, but adults can live for several years without feeding.

Behaviour
They actively seek hosts by crawling short distances from their resting places at night. A blood-meal is taken within about 30 minutes.

Disease
HUMANS *Ornithodoros savignyi* causes biting stress and toxic or allergic effects leading to acute illness or paralysis. Species in the *O. moubata* complex transmit the spirochaete bacterium *Borrelia duttoni* causing tick borne relapsing fever.

DOMESTIC ANIMALS Heavy infestations of pig houses lead to loss of productivity of the pigs. *Ornithodoros savignyi* causes biting stress to livestock and in some cases the saliva has toxic and allergic effects. Paralysis may be caused by the feeding of *O. savignyi* and *O. lahorensis*. *Ornithodoros moubata* and other species feeding on pigs transmit the virus causing African swine fever between pigs and between warthogs (this virus is also transmitted contagiously).

2.1.3 *Otobius* soft ticks

This is a genus of two species.

Structure

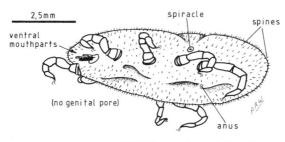

Figure 2.8

OTOBIUS second nymph, lateral

- *Otobius megnini* is known as the spinose ear tick because there are thick sharp bristles like spines all over the body of the nymph.
- Other features are typical of soft ticks.
- The shape of the nymph from above is oval with a constriction at the posterior half of the body; but there is no lateral suture.
- The integument of the nymph does not have the flat areas or bumps (mamillae) characteristic of *Argas* or *Ornithodoros*.
- In adults the integument is textured more like that of other soft ticks and the hypostome is poorly developed.

Sexes
These are similar except for small differences in the genital pore.

Distribution
Otobius megnini and *O. lagophilus*, originated in Nearctic and Neotropical Regions but *O. megnini* has spread into some southern and eastern areas of the Afrotropical Region, and into the Oriental Region. It is also found in some cool humid climates (in Canada).

Hosts
Otobius megnini feeds in the ears of cattle, horses, sheep, deer, dogs and other mammals. It has been recorded feeding on humans.

Life cycle
This is a one-host type but is unusual for ticks because the adults do not feed. The females lay eggs in the soil or cracks in animal housing, larvae crawl onto hosts and attach and feed in the ears where they then moult

and go through two nymphal stages. The nymphs feed for weeks to several months in the ears. They then detach and moult off the host into adults. These reproduce on the ground.

Behaviour
Only adults and larvae are found off the host.

Disease
HUMANS Infestation of the ear by nymphs may occur.

DOMESTIC ANIMALS *Otobius megnini* causes biting stress and damage to cattle and other animals when in heavy infestations. Damage to ears may permit infestation by larvae of flies, causing myiasis.

2.1.4 *Boophilus* hard ticks
A genus of five specialized species, known as cattle ticks or blue ticks.

Structure
- Unfed ticks are small and pale coloured.
- There are no pigmented patterns on the integument, but it is thin enough for the dark pattern of the gut to show through in unfed ticks.
- The mouthparts are very short and the palps have an appearance of two sections formed as protruding rings. However the second segment does not extend beyond the margin of the basis capituli as in *Haemaphysalis* ticks.
- Eyes are present but small.
- The anal groove is posterior to the anus but is shallow and faint.
- The first coxa is without long spurs.
- In males the anal plates are well developed.
- Important features for identifying species are as follows: The length of spurs on the first coxa, the number of columns of teeth on the hypostome of the mouthparts, and the presence of a spur with setae on the first palp segment (Figure 2.9g). The hypostomal teeth are difficult to see. If the mouthparts are removed these teeth can be seen under a compound microscope, and they shine brightly with ultraviolet light using a fluorescence microscope. These teeth occur in columns along the length of the ventral side of the hypostome. The hypostome forms one part of the tube which is inserted into the host, the cheliceral sheath forms the other part.

A full key should be used to identify Boophilus to species, the species in Figure 2.9 are examples only.

INTRODUCTION TO SOFT AND HARD TICKS

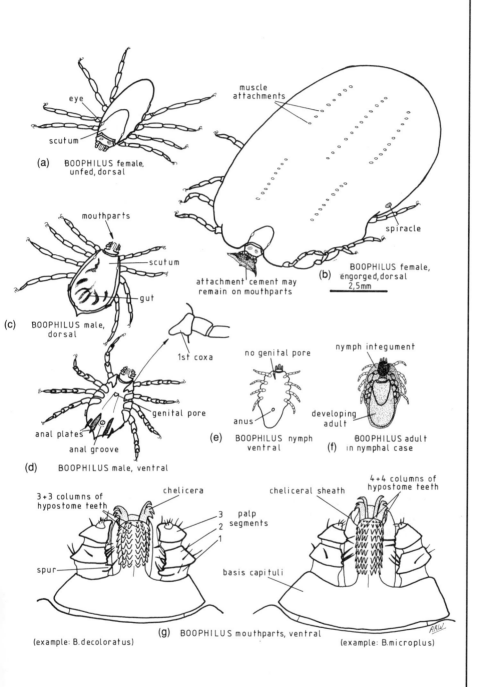

Figure 2.9

Sexes
Males have a large scutum and fed males have anal plates and **caudal** appendage protruding.

Distribution
Originally they occurred in Oriental and Afrotropical Regions but have spread with cattle movements, particularly *B. microplus* which is common in tropical humid climates of the Neotropical and Australasian Regions and is present in the Afrotropical Region.

Hosts
Boophilus ticks are found mainly on cattle. Small numbers may be found on other livestock such as horses and sheep, and on wild ruminants. Ticks on horses that look like *Boophilus* may be *Anocentor (Dermacentor) nitens* or *Margaropus winthemi*.

Life cycle
This is always the one-host type (Figure 2.4).

Behaviour
This is typical of one-host ticks, larvae are the only stage that seek hosts and they cluster at the top of grass stems waiting for a host to pass. On cattle larvae attach over wide areas of the back, neck and belly. The adult ticks need to change position for mating but remain widely distributed on the host.

Disease
DOMESTIC ANIMALS Heavy infestations cause biting stress, loss of production and damage to valuable areas of hide. They transmit protozoa causing babesiosis. Both *Boophilus microplus* and *B. annulatus* transmit both *Babesia bigemina* and *B. bovis*. *Boophilus decoloratus* transmits *Babesia bigemina*. They are one of the factors in the transmission of the rickettsias *Anaplasma centrale* and *A. marginale* causing anaplasmosis in cattle.

2.1.5 *Amblyomma* hard ticks

A genus of about 100 species which vary greatly in size and preferred hosts; in southern Africa they are known as bont ticks.

Structure

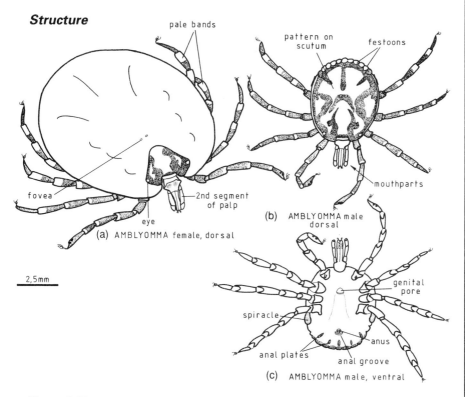

Figure 2.10

- Mouthparts are longer than the basis capituli, and have an elongated second segment of the palps.
- The scutum of adults usually has a bright pattern of various coloured pigments.
- *Amblyomma americanum* is known as the lone star tick because of a single pale spot on the scutum.
- Legs have white pigment in rings at the far end of each segment (compare with *Hyalomma*).
- Eyes are present.
- The anal groove is conspicuous and posterior to the anus.
- Festoons are present.
- Anal plates in males may be present but these are usually small.
- Spurs on the first coxae may not be obvious.

38 TICKS

Sexes
The coloured pattern is most conspicuous in males because of their larger scutum.

Distribution
These ticks are widespread throughout tropical humid climates and some parts of warm humid climates. Thus they have a wide zoogeographical range, but are less common in Nearctic, Palaearctic and Australasian Regions. *Amblyomma variegatum* has spread with cattle movements from Africa to Caribbean islands.

Hosts
Amblyomma feed on a wide variety of mammals, birds and reptiles, and sometimes amphibians. Cattle, sheep and goats are parasitized by numerous species of *Amblyomma*, and larvae of such species may feed on small mammals or birds. Humans may be seriously infested with larvae from cattle pastures.

Life cycle
All species have a three-host cycle (Figure 2.3).

Behaviour
This is typical of hard ticks, larvae occur in clusters at the top of grass stems, nymphs and adults usually occur singly on the ground but some species occur on vegetation. Favourite sites for feeding on domestic stock are udder, belly, groin, axilla and dewlap. Males may remain at the same site for months, attracting females to attach at the same site and then mate.

Disease
HUMANS *Amblyomma cayennense* and *A. americanum* transmit *Rickettsia rickettsii* causing Rocky Mountain spotted fever. African species of *Amblyomma* transmit *Rickettsia conori* causing tick typhus.

DOMESTIC ANIMALS Heavy infestations of species feeding on livestock, for example *A. variegatum* and *A. hebraeum*, can lead to production losses by reduction of milk and weight gain, and by damage to udders. *Amblyomma variegatum, A. hebraeum* and other species transmit the rickettsia *Cowdria ruminantium*, the causative organism of heartwater. *Amblyomma variegatum, A. hebraeum* and several other species transmit the protozoan *Theileria mutans* to cattle. The bacterial skin disease dermatophilosis is made much worse when *Amblyomma variegatum* adults feed on cattle.

2.1.6 *Hyalomma* hard ticks

This is a genus of about 30 species, known in southern Africa as the bont legged ticks.

Structure

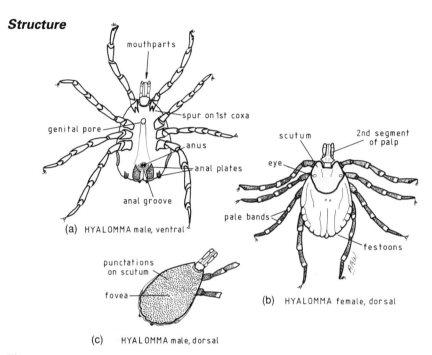

Figure 2.11

- Mouthparts are longer than the basis capituli, with an elongated second palp segment.
- The scutum in both sexes is a plain dark brown colour without pigmented patterns (but *H. albiparmatum* has a white parma). It may have patterns formed by many hollow pits (punctations).
- Legs have rings of white pigment at the far ends of the segments (but *H. detritum* has only small white patches on the legs).
- The coxae of the first legs are divided by two large spurs.
- Eyes are present.
- The anal groove is conspicuous and posterior to the anus.
- Anal plates of the males are well developed.
- Festoons are present but often are not well defined.

Sexes
Males have a large scutum and fed males have protruding anal plates.

40 TICKS

Distribution
Hyalomma ticks are found in tropical humid and dry climates in the Afrotropical, Palaearctic and Oriental zoogeographical Regions. They are adapted to dry environments, and in areas of little vegetation they live around mammal nests or livestock housing.

Hosts
They feed on a wide variety of mammals including cattle, camels, sheep and goats. Immature stages will feed on small mammals such as gerbils and hares; *Hyalomma aegyptium* is often found on pet tortoises.

Life cycle
Hyalomma ticks have one-, two- or three-host cycles and some species vary between two- and three-host cycles depending on the species of host. Timing of stages in the life cycle, particularly of moulting nymphs, is adapted to survival during cold dry winters.

Behaviour
Adaptations to dry habitats include actively walking over ground from sheltered resting sites to find hosts which rest nearby. In some species moulting larvae and nymphs hide in cracks and crevices within housing of livestock. Preferred feeding sites are on the belly, groin, axillae, perianal region, feet and tip of the tail.

Disease
HUMANS Several species transmit the virus causing Crimean–Congo haemorrhagic fever, and several species transmit *Rickettsia conori* causing tick typhus. The rickettsia *Coxiella burneti* causing Q-fever may be transmitted by *Hyalomma* ticks but is commonly transmitted contagiously.

DOMESTIC ANIMALS Feeding on feet or perianal region may lead to disablement. The salivation of some species (e.g. *Hyalomma truncatum*) causes a toxic disease known as exudative dermatitis or sweating sickness. Paralysis may be caused by the same species. The rickettsia *Ehrlichia bovis* causing bovine ehrlichiosis is transmitted by *Hyalomma*. Several *Hyalomma* species transmit the *Babesia* species causing babesiosis in horses and dogs. Many *Hyalomma* species can transmit several *Theileria* species, for example *T. annulata* which causes tropical theileriosis in cattle.

2.1.7 *Rhipicephalus* hard ticks

This is a genus of about 70 species, most of them confined to Africa but often very numerous there.

Structure

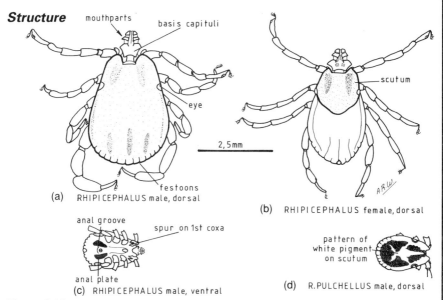

Figure 2.12

- Mouthparts are short (not longer than the basis capituli), and the basis capituli has a six-sided shape.
- The first coxa is divided into large spurs.
- Eyes are present.
- Festoons are present.
- The anal groove is obvious and posterior to the anus.
- The scutum may have punctations and grooves forming patterns.
- The anal plates in males are conspicuous.
- There are usually no coloured patterns but the common *R. pulchellus* has a pattern of white pigment on the scutum and is known as the zebra tick. The rarer *R. humeralis, R. maculatum* and *R. dux* have white patterns.

Sexes

Males often have the fourth legs much thicker than in females. Males have a large scutum and fed males have the anal plates standing out and the parma (central festoon) may protrude as a caudal appendage.

Distribution
Most species occur in the Afrotropical Region, a few in the tropical and warm humid climates of the Oriental and Palaearctic Region. The dog tick or kennel tick *R. sanguineus* has spread far with dogs and may infest heated kennels in cool climates. *Rhipicephalus sanguineus* is in a species group of several very similar species.

Hosts
The usual hosts are wild bovid species, cattle, sheep, goats, and a wide variety of small mammals. The important species *R. appendiculatus* (the brown ear tick of East and southern Africa) has adapted to the presence of cattle and often the entire life cycle is supported by cattle.

Life cycle
They are typical three-host ticks (Figure 2.3), but *R. evertsi* (the red legged tick) has a two-host cycle on livestock where the larvae and nymphs infest the ears. *Rhipicephalus bursa* is another two-host species feeding on livestock.

Behaviour
Most species require humid microclimates and some (for example *R. appendiculatus*) are best adapted to a cool climate at high altitudes on the African plateau, but may also occur near the coast where the humidity is high enough.

Disease
HUMANS Several species transmit *Rickettsia conori* causing tick typhus. The bacterium *Francisella tularensis* causing tularaemia in humans is transmitted by *R. sanguineus*. The larvae of many species cause biting nuisance.

DOMESTIC ANIMALS *Rhipicephalus appendiculatus* is notorious as the vector of *Theileria parva* causing East Coast fever in cattle. *Rhipicephalus appendiculatus* and *R. pulchellus* transmit *Theileria taurotragi* to cattle. Several species transmit *Ehrlichia* species of rickettsia causing ehrlichiosis in cattle, sheep and dogs (in the latter called canine pancytopenia); Nairobi sheep disease virus; and the protozoans *Babesia equi* to horses and *B. canis* to dogs; the latter causes malignant jaundice. Heavy infestations of *Rhipicephalus* ticks cause loss of gain in liveweight of cattle and tick toxicosis. Paralysis of livestock is sometimes caused by *R. evertsi* and *R. bursa*. Species infesting ears may occur in such high numbers that they cause wounds which become infested with screwworm (*Chrysomya* fly larvae, Calliphoridae) thus leading to myiasis.

2.1.8 *Dermacentor* and *Anocentor* hard ticks

Dermacentor is a genus of about 30 widespread species. Included here is *Anocentor nitens*, sometimes included in the genus *Dermacentor* as *D. nitens*.

Structure

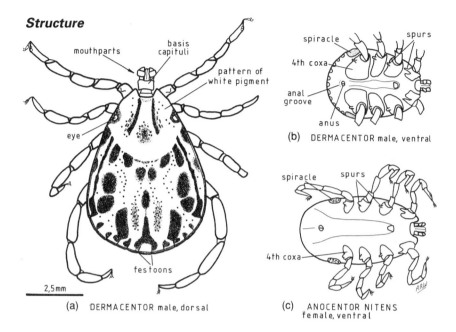

Figure 2.13

- These are large ticks with complex patterns of white pigment on the scutum.
- Mouthparts are fairly short (not much longer than basis capituli), and the basis capituli has a rectangular shape.
- Eyes are present.
- There are eleven festoons.
- The first coxa is divided into spurs and there may be spurs on the other coxae.
- There are no anal plates and the anal groove is inconspicuous.
- The spiracle is shaped like a comma.
- *Anocentor nitens* (Figure 2.13c) is a small tick without a white pattern on the scutum. It may be confused with *Boophilus* but it is darker and has larger spurs on the first coxa and small spurs on the other coxae, eyes are present but small, and there are seven festoons. The spiracle is circular with an inner ring of round spots. Males of *A. nitens* have the large fourth coxae and lack of anal plates characteristic of *Dermacentor* species.

Sexes
The white pattern is most conspicuous in males because of the large scutum; males usually have large coxae, particularly the fourth pair.

Distribution
They are widely distributed in all regions except Australasian but are most common and of signifance to humans and domestic animals in the Nearctic Region. The two Afrotropical species infest elephant and rhinoceros.

Hosts
Dermacentor species feed on a wide variety of mammals including large wild ungulates and livestock. Immature stages may feed exclusively or mainly on small mammals. *Dermacentor andersoni* and *D. variabilis* adults will infest humans. *Anocentor nitens* feeds on horses.

Life cycle
Most species have a three-host cycle but *D. albipictus* and *D. nigrolineatus* have a one-host cycle on deer, cattle and horses, and *A. nitens* has a one-host cycle on horses.

Behaviour
They require humid microclimates for survival of questing stages.

Disease
HUMANS Several *Dermacentor* species may cause paralysis. *Dermacentor* species transmit the viruses causing: Russian Spring–Summer encephalitis, tick borne encephalitis and Colorado tick fever. Rickettsias transmitted are *Rickettsia rickettsii* causing Rocky Mountain spotted fever (Mexican spotted fever, tick typhus) and *R. sibirica* causing Siberian tick typhus. The bacterium *Francisella tularensis* causing tularaemia may be transmitted by *Dermacentor* species. The rickettsia *Coxiella burneti* causing Q-fever may be transmitted by *Dermacentor* ticks but is commonly transmitted contagiously.

DOMESTIC ANIMALS *Dermacentor* species may cause paralysis in cattle and pet animals. *Dermacentor albipictus* may build up to massive infestations on horses, deer or moose, causing stress or death. The rickettsia *Anaplasma marginale* causing anaplasmosis in cattle is transmitted by several *Dermacentor* species. The *Babesia* species causing babesiosis in horses are transmitted by *A. nitens* and *Dermacentor* species, and those in dogs by several *Dermacentor* species.

2.1.9 *Haemaphysalis* hard ticks

Haemaphysalis is a genus of about 150 species.

Structure

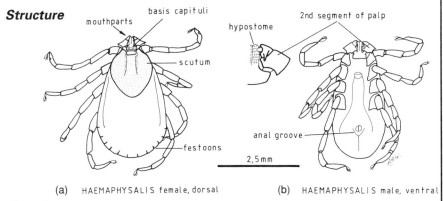

(a) HAEMAPHYSALIS female, dorsal   (b) HAEMAPHYSALIS male, ventral

Figure 2.14

- These are small and pale coloured ticks without pigmented patterns.
- Mouthparts are short with the second segment of the palps extended at the outer edge, (in some species this segment merges with the outline of the first legs or the scutum forming a cone shape).
- The basis capituli is of rectangular shape.
- The coxa of the first leg may have a small spur but is not divided, the other coxae may have spurs.
- The anal groove is posterior to the anus.
- There are eleven festoons.
- There are no eyes.
- There are no anal plates.

Sexes
There is little difference in the sexes except for the usual large scutum of the male.

Distribution
They are found in all climatic zones and zoogeographical regions but most species occur in the Oriental Region. Humid microclimates are usually preferred.

Hosts
A wide variety of mammals is parasitized, with birds sometimes used as hosts. A small number of species are normal parasites of livestock and a

common pest of dogs in tropical countries is *Haemaphysalis leachi*, the yellow dog tick.

Life cycle
A three host cycle is normal.

Behaviour
Haemaphysalis ticks tend to be closely associated with nests of hosts but may also occur in open pasture.

Disease
HUMANS *Haemaphysalis* species transmit the virus causing Russian Spring – Summer encephalitis, tick borne encephalitis and Kyasanur forest disease. The bacterium *Francisella tularensis* causing tularaemia is transmitted by *Haemaphysalis leporispalustris*. *Rickettsia sibirica* and *R. conori* causing tick typhus are transmitted by several *Haemaphysalis* species.

DOMESTIC ANIMALS The protozoan *Babesia canis* causing malignant jaundice in dogs is transmitted by *H. leachi*. *Babesia major*, causing a benign babesiosis of cattle, is transmitted by *H. punctata*.

2.1.10 *Ixodes* hard ticks

Ixodes is the largest tick genus, with about 220 species spread throughout the world. *Ixodes* is a genus by itself in a group known as the Prostriata, the other genera of hard ticks are in the Metastriata.

Structure
- The mouthparts of the females are long, with the second palp longer than the others and shaped with a gap between it and the hypostome.
- The anal groove is conspicuous and anterior to the anus (in all other hard tick genera the anal groove is posterior to the anus).
- The coxae, particularly the first, may have long single spurs but are not divided.
- There are no eyes.
- There are no festoons.
- The colour of the integument of unfed *Ixodes* may be dark or reddish brown, but is without pigmented patterns.

Sexes
Males are smaller than females and mouthparts even smaller in propor-

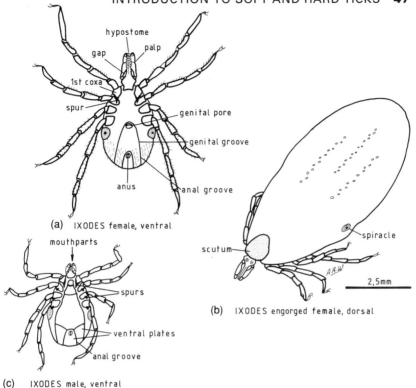

Figure 2.15

tion. They have ventral plates as large flat hardened areas covering the ventral surface. The scutum is large as in other ticks. The distinct anal groove appears as the boundary of the anal plate.

Distribution
They are found throughout all zoogeographical regions and in a wide variety of climates but usually require wetter microclimates than other genera of ticks.

Hosts
The variety of hosts is very wide. Some species are very specialized for certain hosts and habitats (for example bats). Other species will feed on many species of host, particularly small mammals or birds. In cool humid climates this genus is often the commonest tick and is often found on cattle, sheep and goats. *Ixodes ricinus* will feed on cattle, sheep, horse, deer, hares and other small mammals and birds and will readily start feeding on unusual hosts such as humans. This is characteristic of other species important in disease, see below.

Life cycle
This is a three-host type, but the males often mate before the females attach to the host.

Behaviour
Many *Ixodes* species are highly sensitive to drying out, thus their behaviour on vegetation is affected by the need to conserve or regain moisture from humid microclimates.

Disease
HUMANS Feeding often causes swelling around mouthparts and when the tick is removed the mouthparts remain in the skin causing an abscess. Some species cause nausea and fever (*I. pacificus*) others cause paralysis (*I. rubicundus* in southern Africa, *I. holocyclus* in Australia). *Ixodes* species transmit the viruses causing: Russian Spring–Summer encephalitis, tick borne encephalitis, Omsk haemorrhagic fever and Powassan encephalitis virus. *Ixodes* species transmit *Rickettsia conori* and *R. australis* causing tick typhus. *Ixodes* species transmit the bacteria *Francisella tularensis* causing tularaemia, and *Borrelia burgdorferi* causing Lyme disease. The protozoa *Babesia microti* and *B. divergens* normally complete their cycle in other mammals but transmission to humans by *Ixodes* may result in human babesiosis.

DOMESTIC ANIMALS The species of *Ixodes* mentioned above causing paralysis in humans also cause paralysis in a wide variety of livestock species and in dogs. In Europe, *Ixodes ricinus* transmits the following: the virus causing louping ill to sheep, the rickettsia *Cytocetes phagocytophila* causing tick borne fever, the protozoan *Babesia bovis* (one of several babesias causing cattle babesiosis). The feeding sites of *Ixodes ricinus* on sheep may become a site of infection with *Staphylococcus* bacteria causing pyaemia.

2.1.11 Other tick genera
There are other genera which may not be seen often; those which may be found on domestic or zoo animals are as follows:

Antricola
Antricola is a soft tick genus which feeds on bats.

Aponomma
Aponomma is a hard tick genus commonly found on snakes and lizards, they are related to and similar to *Amblyomma* ticks but have no eyes.

Cosmiomma
Cosmiomma is a hard tick genus found on hippopotamus.

Margaropus
A hard tick genus; *M. winthemi* is a one-host tick which feeds in wintertime on horses in dry cool regions of South Africa. It is similar to *Boophilus*, with eyes and short mouthparts but the palps are not ridged. The body colour is dark and there are many setae. The male has a pointed posterior body and very stout leg segments.

Nosomma
Nosomma is a hard tick genus of one rare species, *N. monstrosum*, which feeds on buffalo, cattle and small mammals in the Oriental Region. Eyes are present, the scutum has coloured patterns, the third segment of the palps has a dorsal and ventral extension, the males have anal plates.

Rhipicentor
A hard tick genus similar to *Rhipicephalus*; it is found sometimes on cattle in central Africa. The body is large, eyes are present, mouthparts are short, the basis capituli has very pointed lateral angles, males have very large fourth coxae but have no anal plates.

Part Two
THE INSECTS – INSECTA

THEIR BIOLOGY AS AN AID TO IDENTIFICATION

The insects are separated from the mites and ticks by having the body divided into a head, thorax and abdomen. The thorax has three pairs of legs and may have one or two pairs of wings. The mouthparts are also different but this is not a feature used in this book.

However, there are many complications with insect structure. Some groups of insects (in the sub-class Apterygota) are wingless (apterous). The other sub-class of insects, the Pterygota have wings, or have lost them during evolution (for example, fleas).

The Pterygota insects are in two divisions. One, the Exopterygota, are so named because the wings develop externally during stages of the life cycle so that the immature stages resemble the adults (the Pterygota are also known as the Hemimetabola). These insects go through only a partial (incomplete) metamorphosis between immature and adult stages. The true bugs (Hemiptera) are typical exopterygotes. They include the triatomine bugs and bed bugs. Lice are exopterygotes which have lost their wings in adaptation to a parasitic mode of life in all stages of the life cycle.

The other division, the Endopterygota, are so named because the wings develop internally during a complete metamorphosis from the larval stage to the completely different adult (the Endopterygota are also known as the Holometabola). There is a transitional pupal stage between larva and adult. The true flies (Diptera) are typical endopterygote insects. They include many types of importance to health such as mosquitoes and tsetse. The fleas (Siphonaptera) are endopterygote insects which have lost their wings in adaptation to parasitism during the adult stage. However, the larvae are free-living on debris and similar to other larvae of endopterygotes.

3 FLIES – DIPTERA

3.1 GENERAL INTRODUCTION TO THE FLIES

The true flies (of the order Diptera), have many species parasitic on humans and livestock. The Diptera have a unique feature; the name meaning two-winged. The pair of wings at the front are developed for flight, with the pair of wings at the back being modified to small stalks which are bulbous at the end (Figure 3.1a). These are the halteres, or balancing organs and are essential in flight. When the fly is killed the halter may be fixed in its upper position under the wing where it is difficult to see. Look under the wing to see the halter. In addition many flies have flaps at the base of each wing (one alula and two squamae). These can be mistaken for a second pair of wings, but unlike the wings they have no veins (Figure 3.1b). The squamae may also be called calypters. Some other insects such as bees and wasps of the order Hymenoptera (for example *Vespula*, Figure 3.2) appear to have one pair of wings but the fore- and hindwings are joined in flight by a row of hooks.

The life cycle has a complete metamorphosis (Figures 3.3 and 3.4). The larvae have no legs or wings and they develop through usually three or four stages or instars of similar shape. There may be more stages in some families such as Simuliidae, Tabanidae. They then change into a pupa. In mosquitoes (Figure 3.3) and related flies the pupa remains mobile but does not feed. In *Musca* houseflies (Figure 3.4) and similar flies the pupa develops inside an immobile puparium formed from the last larval skin. The fully formed adult emerges from the pupa. *Glossina* tsetse flies (Figure 3.5) and hippoboscid flies develop one egg at a time within the female so that the larva develops fully in the abdomen of the female. The larva is laid on the ground or at the host's nest by the female and it then forms a pupa inside a puparium. The adult flies mate and feed.

Larvae of mosquitoes and related flies feed and grow on a variety of material in their habitat in water or soil. Larvae of flies with a *Musca* type

54 FLIES – DIPTERA

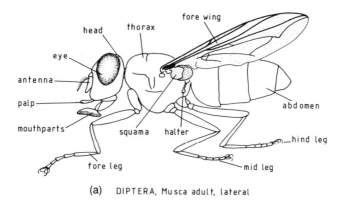

(a) DIPTERA, Musca adult, lateral

Figure 3.1

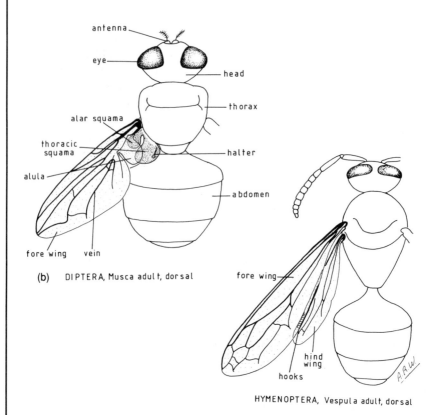

(b) DIPTERA, Musca adult, dorsal

HYMENOPTERA, Vespula adult, dorsal

Figure 3.2

GENERAL INTRODUCTION TO THE FLIES **55**

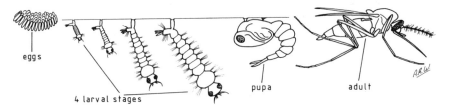

Figure 3.3

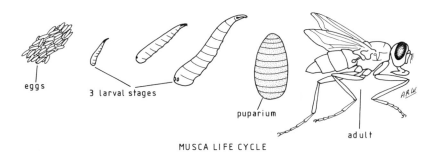

Figure 3.4

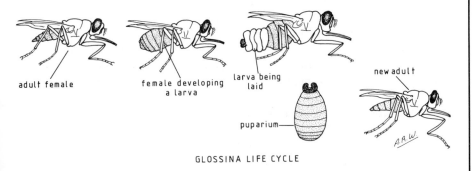

Figure 3.5

of life cycle feed and grow on plant or animal material and some are parasitic such as the flies involved in myiasis. Larvae of *Glossina* and the hippoboscid flies are fed entirely inside their mother by special secretions.

In mosquitoes the females feed on blood and one large blood meal is taken to support production of a batch of eggs. This is repeated several times. In *Musca* and other flies with similar life cycles the female may feed frequently on protein sources to produce egg batches. In *Glossina* both sexes feed on blood and the females take repeated meals to support production of each larva.

The body of adult flies is divided into an obvious head, a thorax with three pairs of legs, and an abdomen with segments that are usually clearly seen. The antennae are important in the identification of main groups of flies (Figures 3.6 to 3.9). The spacing of the eyes is often useful for distinguishing the sexes. Females usually have eyes separated by a wide gap (dichoptic), whereas the eyes of males are usually close together (holoptic). However there are some species where this is the other way round. The mouthparts show great variation for different functions and these do not necessarily correspond with taxonomic groups. Thus the mouthparts may be included in an identification but many other features are more important.

Identification of flies is usually made from examination of the adults, but it may be important to know what larvae are in a habitat for control purposes. Identification keys are available for larvae and pupae of groups such as mosquitoes. The groups of flies having larvae which infest the skin and flesh of animals (a type of parasitism called myiasis), usually need to be identified as larvae. The adult flies are more difficult to obtain for identification unless they are reared through.

For identification of adult flies it is important to establish the sub-orders to which they belong: Nematocera, Brachycera, or Cyclorrhapha. (Note: Cyclorrhapha may be named Muscomorpha and included with Brachycera)

- **Nematocera** (Figure 3.6) have antennae with 11 or more similar segments which are usually long and slender and project forward, but in *Simulium* the antennae are short.
- **Brachycera** (Figure 3.7) have antennae with three larger segments at the base and three or more smaller segments at the far end; the large segments are of different shapes and the antennae project forward from the head.
- **Cyclorrhapha** (or Muscomorpha) (Figure 3.8) have three segments of different shape, the main part of the antenna is the third segment which may be folded down in a groove between the eyes. From the third segment projects a bristle called the arista and this usually has fine setae or hairs on it. The arista may be mistaken for the complete antenna; look at the fly between its eyes to see the main segments.

GENERAL INTRODUCTION TO THE FLIES 57

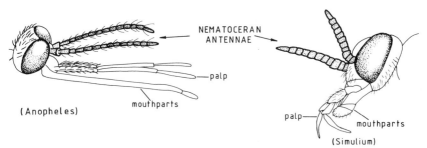

Figure 3.6

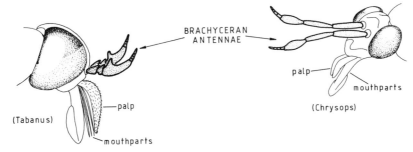

Figure 3.7

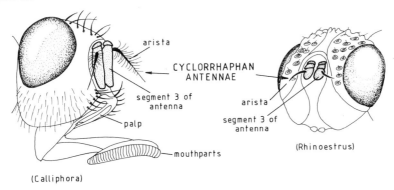

Figure 3.8

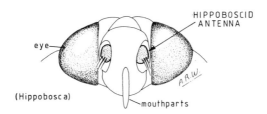

Figure 3.9

58 FLIES – DIPTERA

The hippoboscid flies (Figure 3.9) are usually identified by features other than their antennae, they have reduced antennae as small knobs in pits between the eyes and with several thick setae or bristles protruding.

3.2 NEMATOCERAN FLIES

3.2.1 Phlebotomine sandflies – Psychodidae

The family Psychodidae contains many species. Those important to health are in the genera *Phlebotomus* and *Lutzomyia*, which are in the subfamily Phlebotominae. The name sandfly should not be confused with the same name sometimes given to ceratopogonid midges or simuliid blackflies.

Structure

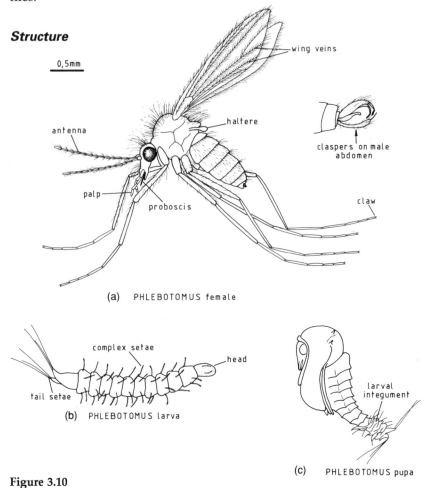

(a) PHLEBOTOMUS female

(b) PHLEBOTOMUS larva

(c) PHLEBOTOMUS pupa

Figure 3.10

- *Phlebotomus* and *Lutzomyia* are very similar in appearance.
- Phlebotomines are nearly as small as ceratopogonid midges but are more like miniature mosquitoes because of their long legs and wings.
- Antennae are of typical nematoceran type and covered in fine setae.
- The palps are long and of five segments.
- The biting mouthparts form a short proboscis.
- The wings are long and pointed and have a simple pattern of veins which are straight along the length of the wing with only three branchings. They are covered in long setae arranged along the veins and at the edges. When the fly is resting the wings are held extended above the thorax.
- The legs are long and the claws at the end are very small.
- The dorsal surface of the thorax is covered in long setae and the abdomen is long and narrow.
- In the sub-family Psychodinae there are common flies in the genera *Psychoda*, *Pericoma* and *Telmatoscopus* which may be confused with plebotomine sandflies. They are often known as mothflies, such as *Psychoda* (Fig 3.11). They do not feed on blood, they are very hairy, the legs are short, the wings have veins with only two branchings and at rest the wings are held folded over the abdomen in an angled shape like a tent roof.

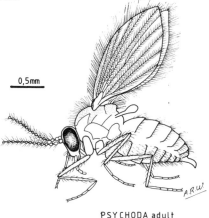

PSYCHODA adult

Figure 3.11

Sexes
The males have prominent claspers at the end of the abdomen, which are used in mating. The female abdomen has small pointed cerci used in egg laying. The antennae and palps of the sexes are similar.

Distribution
Phlebotomus occurs in warm humid and tropical humid climates of the

Afrotropical, Palaearctic and Oriental Regions. Within these regions the flies live in dry areas of savanna and semi-desert vegetation. *Lutzomyia* occurs in tropical humid climates in the Neotropical Region and inhabits wet areas with forest vegetation.

Hosts
Phlebotomus and *Lutzomyia* feed on humans and a wide variety of livestock and other mammals, birds and reptiles. Only the females feed on blood, both sexes feed on plant juices.

Life cycle
The larva develop in a wide variety of diffuse habitats which are cool and moist with decaying debris, but never aquatic. Often these conditions exist in animal burrows, tree holes or termite mounds. Also many species have become adapted to breeding in livestock housing. Pupation occurs in the same site.

Behaviour
The adults fly at night or in cloudy and humid weather. They fly only short distances in a series of very short flights. They rarely occur in large numbers and the flight is without sound so they are not generally noticed when feeding on mammals.

Disease
HUMANS More than 50 species of *Phlebotomus* and *Lutzomyia* transmit species of protozoans causing various forms of leishmaniasis. Species of Protozoa in the complex *Leishmania tropica* are transmitted typically by *P. papatasi*, *P. longipes*, *P. sergenti*, *P. caucasicus* and other species in Africa and Asia. These Protozoa cause cutaneous leishmaniasis (oriental sore). Species in the *Leishmania donovani* complex, causing visceral leishmaniasis (kala-azar), are transmitted typically by *P. argentipes*, *P. chinensis*, *P. martini*, *P. major* and others in Asia and Africa. Species in the *L. donovani* complex, causing visceral leishmaniasis are transmitted by *Lutzomyia longipalpis* in South America. Species in the complexes *Leishmania mexicana* and *L. braziliensis* causing cutaneous and nasal leishmaniasis in Central and South America are transmitted by many species of *Lutzomyia*.
- The bacterium *Bartonella bacilliformis*, causing Carrion's disease is transmitted in South America by *Lutzomyia verrucarum*. This disease is also known as bartonellosis, with two clinical forms, Oroya fever and verruga peruana.
- The virus causing sandfly fever or papatasi fever in Europe, North

and Asia is transmitted mainly by *Phlebotomus papatasi* and *P. sergenti*.

DOMESTIC ANIMALS Many mammals are involved in the transmission cycle of leishmaniasis, of particular importance are dogs, which suffer visceral leishmaniasis.

3.2.2 Biting midges – Ceratopogonidae

The three genera of Ceratopogonidae that feed on humans and livestock (*Culicoides, Forcipomyia, Leptoconops*) are known as biting midges, punkies, sandflies and other names including no-see-ums in the USA. Confusingly, phlebotomine flies are also known (correctly) as sandflies. There are many other types of ceratopogonids that do not feed on humans or livestock. The blood feeding species of the genus *Forcipomyia* may be classified in the sub-genus *Lasiohelea*. The family Chironomidae are often known as non-biting midges or gnats.

Structure
- All ceratopogonids are very small.
- The mouthparts of blood feeding ceratopogonids are complex with six cutting and piercing elements.
- The antennae are long, narrow, with 11 to 15 segments.
- The legs are short.
- The thorax has indentations known as humeral pits.
- Wings have a simple layout of veins and may have a thin covering of short setae (microtrichiae) and also longer setae (macrotrichiae). The radial cells of the wings are typical of ceratopogonids. Such cells are small areas of the wing enclosed by veins, the radial veins in this case. At rest the wings are held over the abdomen.
- Larvae are long and thin with a simple head.
- Pupae are like miniature pupae of mosquitoes, complete with air breathing tubes.

Culicoides
- Species of this genus usually have distinct patterns on their wings.
- The r-m cross vein is present, two radial cells of medium size are present, and the wings are covered in short setae or microtrichiae.
- The antennae of females have 14 or 15 segments.

62 FLIES – DIPTERA

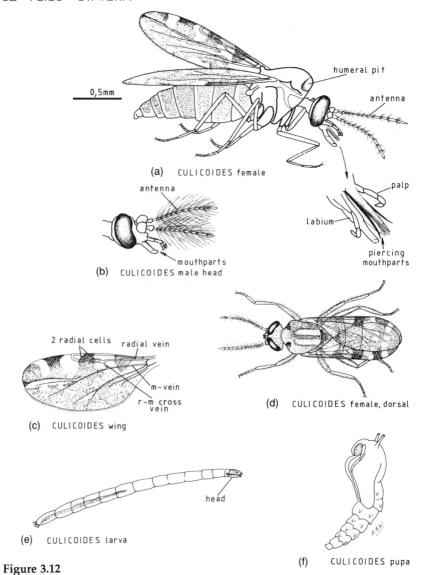

Figure 3.12

Forcipomyia (sub-genus *Lasiohelea*)

- These flies usually lack distinct patterns on their wings.
- The r-m cross vein is present, the wings are covered in long setae or macrotrichiae and microtrichiae are also present, the second radial cell is elongated.
- The antennae of females have 14 or 15 segments.
- There is an obvious empodium between the claws of the feet.

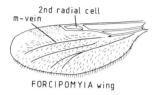

Figure 3.13

Leptoconops

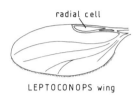

Figure 3.14

- *Leptoconops* species have clear white wings without obvious patterns and without long setae.
- They are without the r-m cross vein, the M vein is not forked and there is often only one radial cell.
- The antennae of females have 11 to 13 segments.

Smaller species of chironomid non-biting midges may be mistaken for ceratopogonid biting midges. They are distinguished by having a very humped thorax, lack of biting mouthparts and different pattern of wing veins without the radial cells.

Sexes
Males have slender abdomens and do not feed on blood, thus mouthparts are less obvious. In males the genital apparatus is more conspicuous than in females and the antennae have long setae.

Distribution
Distribution of ceratopogonids is determined largely by suitable larval habits. These are typically in wet soil between terrestrial and aquatic habitats, or in decaying vegetable material or tree holes. Larvae require wet habitats but are not usually found in open water. The water is usually fresh but some species are adapted to salt water.

Culicoides species occur in all zoogeographical regions and are very common in many habitats in warm humid climates. During summer in cool humid climates *Culicoides* may be extremely numerous in areas with extensive larval habitats. The larvae of *Culicoides* species occur in wet sand and mixtures of wet sand and mud along the sea coast, in lakes and in mangrove swamps. They also occur in peat bogs, temporary pools and mixtures of mud and dung.

Leptoconops species are usually found in tropical humid climates in most zoogeographical regions. Larvae occur in wet sandy or clay soils, often fairly deep down in cracks. They are also common in coastal sandy soils.

Forcipomyia (sub-genus *Lasiohelea*) are found mostly in rainforests (in tropical humid climates) although an important species (*L. sibirica*) occurs in Siberia (cool humid).

Hosts
Ceratopogonids feed on a wide variety of mammals including humans, cattle, sheep, horses and also on birds.

Life cycle
This is of the mosquito type (Figure 3.3). Eggs are laid at the larval habitat. The larvae feed on organic detritus or other small animals; they may develop within a few weeks in tropical climates or take one year in warm or cool humid climates. Pupae live at the surface of the larval habitat, breathing air. The pupal period is short. Only females feed on blood but both sexes feed on plant sugars.

Behaviour
Ceratopogonids do not fly far. They may be dispersed long distances by strong winds but usually the biting adults are found close to the larval habitats. Females are attracted by the smell and warmth of their hosts. The midges land on the hair coat then crawl on the skin between the hairs with wings folded to find a feeding site. Feeding takes a few minutes and is usually nocturnal or at dusk and dawn.

Disease
HUMANS Many ceratopogonids cause substantial biting stress to humans. At tropical tourist beaches the small biting flies called sandflies by local people are likely to be ceratopogonids. The filarial worms *Dipetalonema* (= *Acanthocheilonema*) *perstans*, *D. streptocerca* and *Mansonella ozzardi* are transmitted by *Culicoides*.

DOMESTIC ANIMALS Inflammatory and allergic reactions to ceratopogonid bites may be a problem, such as sweet itch or Queensland itch in horses or leptoconops mange in sheep. *Culicoides* species transmit numerous pathogens including African horse sickness virus, bluetongue virus to sheep, ephemeral fever virus to cattle, the filarial worms

Onchocerca to horses and cattle, and several species of protozoa (*Haemoproteus, Leucocytozoon*) to birds.

3.2.3 Blackflies – Simuliidae

The family Simuliidae has about 1600 species in 19 genera and there are over 40 species of importance to health. The classification is complex. The genus *Simulium* contains most species important to health. Other important genera are *Cnephia*, *Prosimulium*, and *Austrosimulium*. These may be known locally as buffalo gnats (in the USA) because of their shape, or turkey gnats because of their feeding preferences.

Structure

- Simuliids are smaller than other blood feeding flies except for the ceratopogonids and phlebotomines.
- They have short legs, stout bodies and few setae.
- Black colouration is typical but some are yellow or orange.
- The antenna is nematoceran type, (Figure 3.6), usually with 11 similar segments. It is very short and without setae.
- The eyes are large.
- The mouthparts are short, with the biting elements hidden in the labium forming the proboscis.
- The palps are long, with five segments.
- The thorax is humped and the abdomen is clearly divided into nine visible segments. On the first segment there is a thickened area (the basal scale) with a fringe of setae.

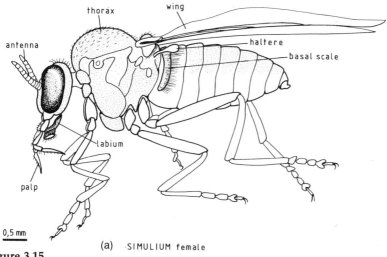

Figure 3.15 (a) SIMULIUM female

66 FLIES – DIPTERA

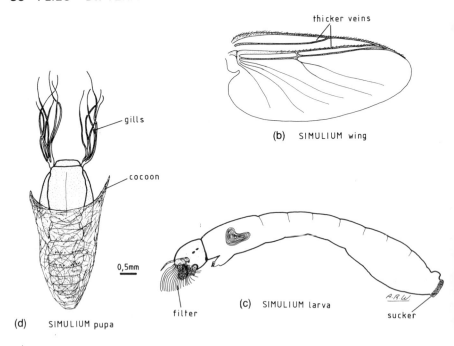

Figure 3.15 *continued*

- The wings are plain, with a simple pattern of veins. The veins are thick at the front of the wing, but they are very faint behind.
- Larvae live in running water. They have a fringe of setae on the head for filter feeding and a 'sucker' (or, more correctly, hook circlet) on the last abdominal segment for attachment to underwater surfaces.
- Pupae remain underwater in a cocoon shaped like a pocket. They have characteristic gills which branch in various shapes depending on species.

Sexes
Male mouthparts are not adapted for blood feeding. Male eyes are larger than those of females, they are close together in the mid-line (holoptic) and with the facets (separate lenses) of the upper part of the eye much larger than those lower down.

Distribution
Simulium occurs in all climates in all zoogeographical regions. The largest numbers of species are found in the Palaearctic Region. *Prosimulium* and *Cnephia* occur in the Palaearctic and Nearctic Regions and *Austrosimulium*

occurs only in the Australasian Region. The larvae are restricted to streams and rivers thus simuliids have a patchy distribution in dry climates. Large populations of simuliids can occur in cool humid climates of northern countries.

Hosts
Many species of simuliids feed as adults on the blood of humans, cattle, horses, sheep, goats, poultry, other livestock and wild mammals and birds.

Life cycle
Eggs are usually laid in batches on objects in streams and rivers. Some species lay eggs loosely over the water surface. Larvae attach to objects under water and feed by filtering material from the flowing water. Pupation occurs at the same site and the adults emerge underwater. Mating occurs in swarms near the larval site. In cool humid climates the emergence of simuliid adults is in the summer and very large numbers may accumulate. Both males and females feed on plant nectar. However, in most species the females need blood for the eggs to develop.

Behaviour
The adults fly and feed in the day-time and are not restricted to shaded or humid sites. They are attracted to hosts from far away by smell and when closer by sight. The female flies crawl on the host to find a feeding site. They cut into the skin by the slashing action of complex mouthparts.

Disease
HUMANS *Simulium* species transmit the filarial nematode *Onchocerca volvulus* which causes human onchocerciasis or river blindness. The commonest species involved are in the complex of species around *S. damnosum* and *S. neavei* in Africa and *S. ochraceum* and *S. metallicum* in Central and South America. Simuliids are often a very severe biting pest, they may occur in dense infestations and cause stress, fever, nausea and allergic dermatitis. The species most often involved are: *Prosimulium mixtum* and *S. venustum* in North America, *S. antillarum* in the Caribbean, *S. indicum* in India and *S. amazonicum* in South America.

DOMESTIC ANIMALS Biting stress to livestock can be very severe, particularly cattle, horses, sheep and goats. Dense infestations occur in northern countries during the summer breeding season. Biting stress is then intense and is complicated by toxaemia and allergic reactions caused by the fly's saliva. This can result in substantial loss of production and numerous sudden deaths in the herds under attack. The important species of biting pest are: *Cnephia pecuarum, Austrosimulium pestilens, S. arcticum, S. colombaschense, S. erythrocephalum* and *S. ornatum*. The protozoan *Leucocytozoon* can be transmitted to poultry by simuliids and *Onchocerca gutturosa* and *O. gibsoni* are transmitted to cattle by *Simulium* species, causing bovine onchocerciasis.

3.2.4 Mosquitoes – Culicidae

The family Culicidae has about 3450 species in 37 genera. Mosquitoes of most importance to human and animal health are in the eight genera included here. Because of their importance in transmission of disease pathogens to humans there are often detailed keys to species of mosquitoes of particular regions, but these are difficult to use without experience. Species of the genus *Culex* are sometimes called gnats.

> *There are many difficulties with identification of mosquitoes and much caution should be used even in identifying the genera.*

General features

These features are illustrated in Figures 3.16, 3.17 and 3.18.

- Mosquitoes are typical Diptera with one pair of wings and one pair of halteres.
- The wings have small squamae, and small scales arranged along the veins of the wing.
- The legs are thin and long and have small claws.
- The complex mouthparts are formed into a long proboscis which projects forwards as a thin tube.
- The antennae are as long as the proboscis, they are typical of the suborder Nematocera (Figure 3.6) with many similar segments.
- The palps may be short or long.

There are three sub-families of mosquitoes: Anophelinae, consisting of *Anopheles* and two other rare genera (*Bironella* and *Chagasia*); Culicinae consisting of nearly all other genera; and Toxorhynchitinae with species that have a sharply bent proboscis and which do not feed on blood.

Other families of Diptera are similar to mosquitoes: the Chaoboridae (phantom midges), Dixidae, Chironomidae (gnats or non-biting midges) and Tipulidae (crane flies). None of these have the long thin mouthparts formed into a projecting proboscis that is so typical of mosquitoes. The larvae of the first three of these families are aquatic but do not have the thoracic segments fused into a large and wide unit.

Diagnostic features of important genera are given in the next section.

Sexes
Males have antennae with numerous long fine setae projecting outwards like a bottle brush. Antennae of females have only short setae. The palps of males are usually about as long as the antennae and proboscis, but in *Haemagogus* and *Sabethes* they are short.

Distribution
Mosquitoes are very widely distributed, in diverse habitats in all climatic zones and all zoogeographical regions. However, they are always dependent on standing water for the larval stages.

Hosts
Female mosquitoes feed on the blood of a wide variety of mammals, birds, reptiles and amphibians. Some species have very distinct host preferences, but species living where hosts are scarce such as in arctic regions will feed on a wide variety of hosts.

Life cycle
Eggs are laid on or near stagnant water and they hatch into larvae which live just under the surface of the water. Larvae feed by filtering material from the water. They pass through four similar stages followed by the pupal stage. The pupa is mobile and remains below the water surface but does not feed. The adults emerge from the pupa and the females seek a host for blood feeding (Figure 3.3). Males feed on plant sugars; females also supplement their blood meals with plant sugars. When the female engorges on blood a batch of eggs is developed. After these are laid the female feeds again to support more batches of eggs.

Behaviour
Most mosquitoes fly at night and the females feed most actively at dusk. Females and males often make a high pitched buzzing noise when flying. When the female has found a host she lands and rapidly inserts the piercing components of the proboscis into the host skin. Blood is taken rapidly, direct from a capillary.

Disease
The most important diseases associated with mosquitoes are listed. There are many other pathogens, particularly viruses, which are also transmitted to humans and domestic animals. Biting stress can be extremely severe and makes some areas at some seasons uninhabitable.

Humans

- **Malaria:** caused by *Plasmodium* protozoa, species infective for humans are transmitted only by *Anopheles* mosquitoes of many species.
- **Filariasis:** caused by species of *Wuchereria* and *Brugia*, transmitted by *Culex, Anopheles, Aedes, Mansonia*.
- **Yellow fever:** caused by a virus, transmitted mainly by *Aedes aegypti*, also by *Haemagogus* and *Sabethes*.
- **Dengue:** caused by a virus; transmitted by *Aedes*.
- **Encephalitis:** caused by viruses; various forms such as Japanese, and St Louis, also the equine forms (Western, Eastern, Venezuelan) infect humans; transmitted by *Culex*.
- **Rift Valley fever:** caused by a virus; transmitted by *Aedes, Culex*, and possibly other genera.
- Other viruses causing clinically recognized diseases and transmitted by *Aedes* and *Culex* are as follows: Chikungunya, O'Nyong-nyong, Ross River.
- **Biting stress:** caused by many species when in large numbers.

Domestic animals:

- **Rift valley fever:** caused by a virus; transmitted by *Aedes* and *Culex*, possibly other genera.
- **Equine encephalitis:** caused by viruses; various forms: Eastern, Western, Venezuelan, transmitted variously by all mosquitoes detailed here.
- **Heartworm of dogs:** caused by *Dirofilaria immitis*; transmitted by *Aedes*.
- **Biting stress:** caused by many species when in large numbers; may be so severe as to cause deaths and make areas uninhabitable by livestock.

Diagnostic features of Mosquito genera

Figures 3.16 and 3.17 compare the general features of anopheline mosquitoes (for example *Anopheles*) and culicine genera (for example *Culex*). Figure 3.18 shows the diagnostic features of *Culex* but is also to show the positions of features for identification of other genera.

- From above the thorax has a rounded scutum with patterns of small setae and scales. At the back of the scutum is the scutellum and postnotum. The scales which may be on the thorax and abdomen have different characteristics between the genera.
- The wing has a characteristic pattern of veins and the squama and alula are always small. The shape of the scales is characteristic and the stem vein may have bristles.
- The claws are small and under high magnification may be possible to see pulvilli and a central empodium as a seta. The claws of the forelegs

NEMATOCERAN FLIES 71

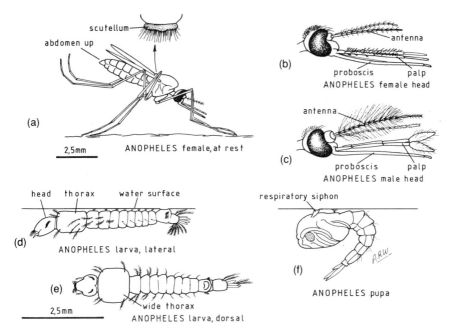

Figure 3.16

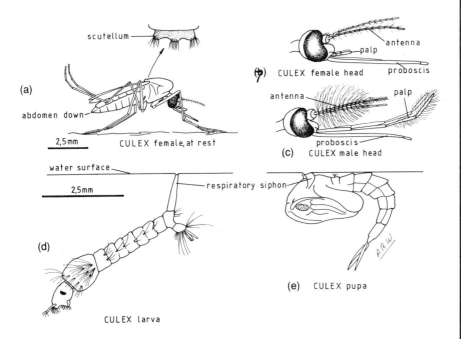

Figure 3.17

72 FLIES – DIPTERA

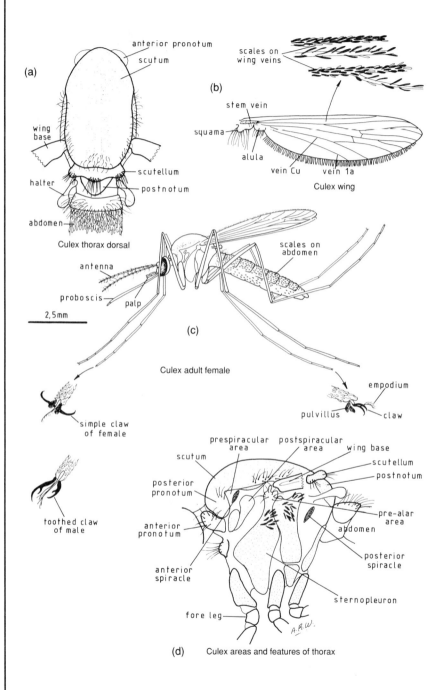

Figure 3.18

may have a branch (toothed claw), usually in the males but sometimes in both sexes.
- The side of the thorax has characteristic patterns of plates and bristles. The anterior pronotum varies in size. The posterior pronotum usually has a group of bristles in front of the anterior spiracles; these are important because they are easily confused with the true prespiracular bristles which occur on the area immediately in front of the spiracle. The postspiracular bristles occur on the plate immediately behind the spiracle. There is also a posterior spiracle below the haltere. The sternopleuron forms into a knob in the pre-alar region just anterior to the wing, this usually has bristles which should not be confused with the postspiracular bristles.

Anopheline genera

Anopheles

Figure 3.16 shows the features of *Anopheles* as the single most important anopheline genus.

- The larvae live just below the water surface in a horizontal position, the head can rotate so that the mouthparts are in contact with the surface of the water for feeding. The spiracles which are in contact with the air do not have a long siphon.
- On the adults the abdomen has few or no scales.
- The females have long thin palps. The long palps of the males are stout at the far end, like a club.
- When *Anopheles* is at rest the abdomen is raised up at an angle to the surface.
- The wings usually have conspicuous patterns of dark and pale patches along the veins.
- The adults have a scutellum with a slightly wavy posterior outline and evenly distributed bristles.
- There are no postnotal bristles.
- *Anopheles* species occur in all zoogeographical regions.

Culicine genera

The remaining genera of importance to health are all in the Culicinae. Figure 3.17 shows *Culex* as an example of culicine mosquitoes. They are all distinguished from *Anopheles* as follows:

- Larvae hang down from the surface of the water, remaining in contact with the surface with the air breathing siphon (they swim down rapidly if disturbed).

74 FLIES – DIPTERA

- The adults usually have a scutellum with the posterior margin having three distinct lobes, each with a separate tuft of bristles.
- The abdomen has a dense uniform covering of scales.
- The females have short palps. The palps of the males are usually long (they are short in *Haemagogus* and *Sabethes*) and are turned up at the far end but are not thickened.
- When at rest the adults have the abdomen bent down at an angle to the surface.
- The wings usually have no distinct patterns, or may have fine patterns of scales of mixed colours.

The seven culicine genera detailed below have the following characteristics in common which help to distinguish them from other culicine genera. Figure 3.18 shows the position of general features.

- The alula has setae or scales and the squama has a fringe of setae.
- The vein 1A (also known as vein 6) reaches the margin of the wing beyond the level of the fork in vein Cu (also known as vein 5).
- The wing scales are narrow (except in *Mansonia* and *Coquillettidia*).
- The postnotum has no bristles (except in *Sabethes*).

Culex

Diagnostic features are illustrated in Figure 3.18 and also in Figure 3.17.

- The scales are narrow and not metallic coloured.
- There are no distinct patterns on the body.
- The abdomen is blunt ended.
- Prespiracular and postspiracular areas are without bristles.
- Claws on the forelegs of the males are toothed, on the females they are simple.
- The feet of females have large pulvilli.
- *Culex* occurs in all zoogeographical regions.

Aedes

- The scales are narrow. There are usually coloured patterns on the body; these may be silvery.
- The abdomen is pointed at the end.
- Prespiracular bristles are absent. Postspiracular bristles are present.
- Both sexes have forelegs with toothed claws. The pulvilli are small or composed of only one hair.
- On the hindlegs the first segment of the tarsus is shorter than the tibia.
- The stem vein (base of subcostal vein) of the wing has bristles on the ventral surface.
- *Aedes* occurs in all zoogeographical regions.

NEMATOCERAN FLIES

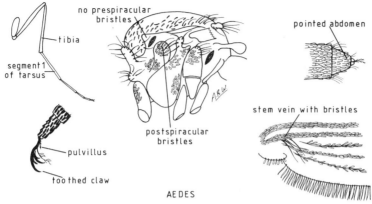

Figure 3.19

Mansonia and *Coquillettidia*

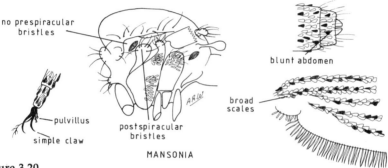

Figure 3.20

- These genera are very similar; sometimes *Coquillettidia* is treated as a sub-genus of *Mansonia*. In *Mansonia* there are postspiracular setae but no prespiracular bristles. *Coquillettidia* has neither prespiracular nor postspiracular bristles and there is an overall yellow colour.
- The body has a covering of broad scales; these may form fine speckled patterns of colour on the wings.
- The abdomen is blunt ended.
- The scales on the wings are very broad.
- The claws are all simple and pulvilli are small or absent.
- *Mansonia* occurs in all zoogeographical regions; *Coquillettidia* occurs mainly in the Afrotropical Region but also in Nearctic, Neotropical, Oriental and Australasian Regions.

76 FLIES – DIPTERA

Haemagogus

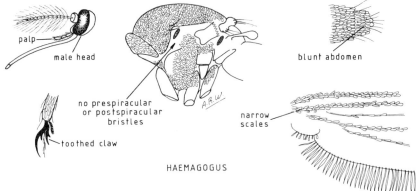

Figure 3.21

- The scutum and abdomen are covered with scales having metallic colours.
- Scales on the wings are narrow; on the body they are fairly broad.
- The anterior pronotal lobes are large and nearly meet at the top.
- The abdomen is blunt ended.
- Both sexes have forelegs with toothed claws.
- There are no bristles in the prespiracular or postspiracular areas.
- The palps of the males are shorter than the proboscis.
- *Haemagogus* occurs in the Neotropical Regions only.

Sabethes

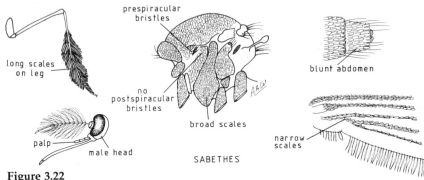

Figure 3.22

- The scutum and abdomen are covered with scales having a metallic colour.
- The scales on the wings are narrow; on the body they are fairly broad.
- The abdomen is blunt ended.
- The anterior pronotal lobes are very large.
- The mid or hindlegs have a conspicuous area of long scales in the shape of a paddle.

- Prespiracular bristles are present but there are no postspiracular bristles.
- There is a group of bristles on the postnotum.
- The palps of males are shorter than the proboscis.
- *Sabethes* occurs in the Neotropical Region only.

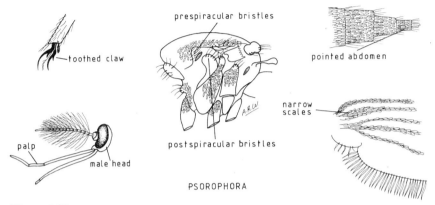

Figure 3.23

Psorophora

- The scales on the body and wings are narrow and without metallic colour.
- Prespiracular and postspiracular bristles are present.
- The abdomen is pointed at the tip.
- Both sexes have forelegs with toothed claws.
- The male palps are longer than the proboscis.
- *Psorophora* occurs only in the Nearctic and Neotropical Regions.

Culiseta

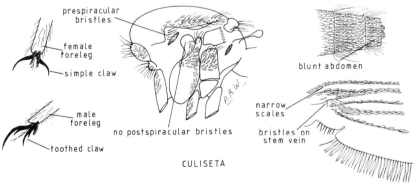

Figure 3.24

- The scales on the body and wings are narrow and not metallic coloured but there are coloured patterns on the body.
- The abdomen is blunt ended.
- Prespiracular bristles are present; postspiracular bristles are absent.
- Forelegs of males have toothed claws, the claws of the forelegs of females are simple.
- The stem vein (base of subcostal vein) of the wing usually has bristles on the ventral surface.
- *Culiseta* occurs in all zoogeographical regions.

3.3 BRACHYCERAN FLIES

3.3.1 Horse flies, deer flies and clegs – Tabanidae

The family Tabanidae has over 30 genera and 4000 species. The genera of most widespread importance to humans and domestic animals are *Tabanus* (horse flies), *Chrysops* (deer flies) and *Haematopota* (clegs). There are genera such as *Lepiselaga, Hybomitra, Silvius, Diachlorus* and many others of importance to health in limited areas. The classification of tabanids is complex and it is not possible here to give details of the numerous other genera with species which may bite humans or domestic animals. The sub-family Pangoniinae contains some species which may be mistaken for ferocious blood feeding flies because the mouthparts form a very long and narrow proboscis that projects forwards. However, this proboscis is used to feed on nectar from flowers, not on blood.

General structure

- The Tabanidae are within the group of flies called Brachycera, with very characteristic antennae (Figure 3.7). The antennae are made up of segments of different sizes and they project in front of the fly (they are often broken in dried museum specimens). There is no arista on the antennae.
- In living specimens the eyes often have brightly coloured patterns. The eyes are generally large and in males they are usually closer together (holoptic) than in females. There may be ocelli (simple eyes) between the main compound eyes near the hind margin of the head.
- The wings are large and may have patterns of dark pigment. There are many strong veins on the wings. These are in a pattern which is fairly uniform in all tabanids but the patterns of pigmented areas are characteristic of genera and species.
- The abdomen has seven segments visible from above.
- The mouthparts are complex. They comprise: a pair of large palps held

out in front of the head; six cutting and piercing elements; a single large labium with a sponging labella. The mouthparts point downwards from the head.
- The feet of the tarsi each have two claws, two pulvilli pads and a central empodium which is formed as a pad, (compare with *Musca*, Figure 3.33).
- The body has numerous fine setae but not the long thick bristles as on muscid and calliphorid flies.
- The larvae are long, cylindrical, with 12 well defined segments, but with a small head. There are rings of tubercles on the segments, these may be known as pseudopods or false legs. There are bands of fine setae around the segments. The posterior of the larvae has a respiratory siphon and a bulge known as Graber's organ.
- The pupae have a rounded anterior end showing the outline of the head. The outlines of the wings and legs are visible. Seven abdominal segments are well defined and moveable and segments 2 to 7 have rings of bristles. The posterior of the pupa ends in a set of spine-like tubercles, known as the caudal aster.

Tabanus

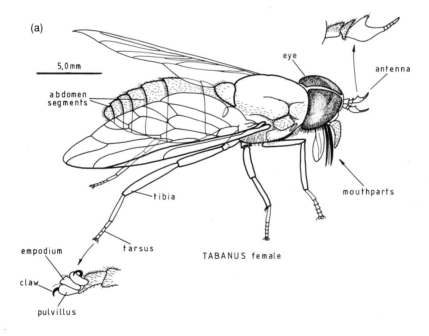

Figure 3.25

80 FLIES – DIPTERA

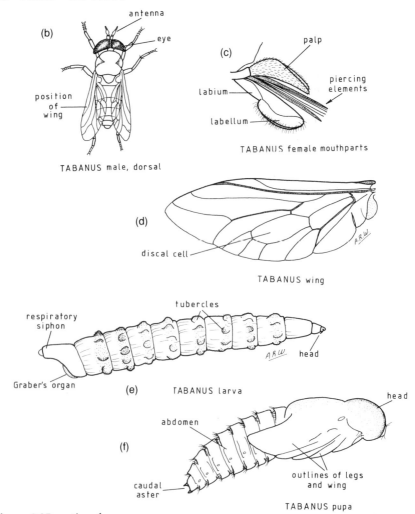

Figure 3.25 *continued*

- These are very large flies.
- The eyes are very large and the gap between them is small even in males. The eyes are green or brown with a horizontal pattern.
- There are no ocelli between the main eyes.
- When at rest the wings are not folded closely over the abdomen. The wings are usually clear but may have a shading of dark pigment without distinct pattern.
- The antennae are short and have seven segments of which the three near the head are large and asymmetrical and the outer four are small and symmetrical.
- There are no spurs on the tibiae of the hindlegs.

BRACHYCERAN FLIES 81

Chrysops

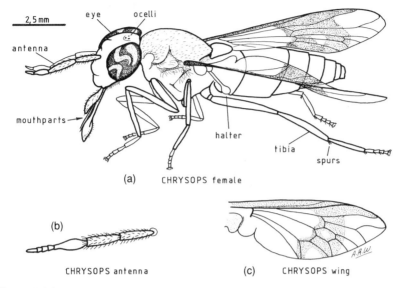

Figure 3.26

- These are medium size flies.
- The eyes are proportionately much smaller than in *Tabanus* and the gap between them is larger. The eyes have red or green patterns in spotted or curved shapes.
- There is a group of three ocelli between the main eyes near the hind margin of the head.
- The wings have a simple pattern of a dark band across the width and when at rest they are held apart over the abdomen.
- The antennae are long with seven segments of which the two near the head are of simple cylindrical shape.
- There are spurs on the tibiae of the hindlegs.

Haematopota

- These are intermediate in size between *Tabanus* and *Chrysops*.
- They have fairly large eyes with a wide gap between them. The eyes have angular patterns of red and green.
- When at rest the wings are held folded close to the abdomen and the wings have complex patterns of dark pigment.
- The antennae are short with six segments of which the three near the head have a curved but symmetrical shape.
- There are no spurs on the tibiae of the hindlegs.

82 FLIES – DIPTERA

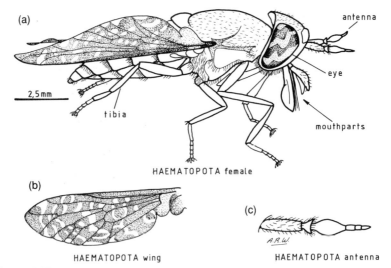

Figure 3.27

Hybomitra

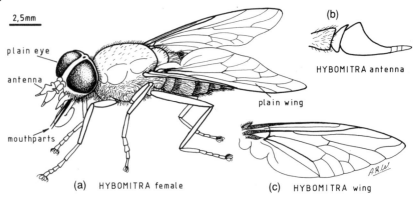

Figure 3.28

- These flies are closely related to *Tabanus*.
- The antenna is similar to *Tabanus* with three large asymmetrical segments near the head and three small symmetrical segments at the outer end.
- The eyes are large and plain.
- The wings are without distinct patterns.
- There are no spurs on the tibiae of any legs.

Lepiselaga

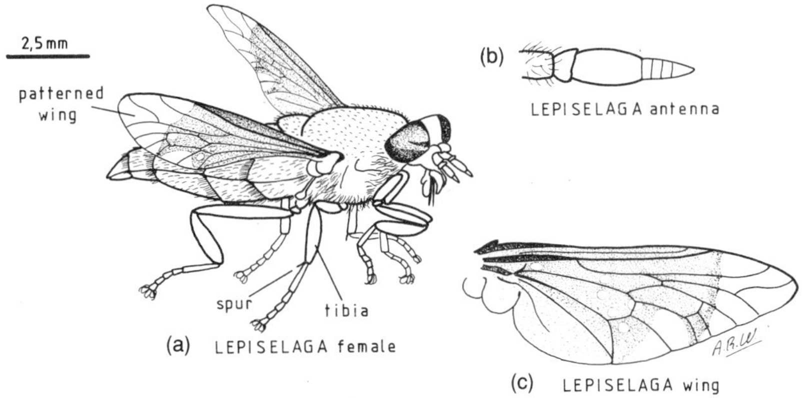

Figure 3.29

(a) LEPISELAGA female — patterned wing, spur, tibia, 2.5 mm
(b) LEPISELAGA antenna
(c) LEPISELAGA wing

- These flies are closely related to *Chrysops*.
- They are smaller than many other typical tabanids.
- The eyes may be patterned and are set far apart.
- The wings have a distinct simple band of colouration.
- There are spurs on the tibiae of the midlegs.

Sexes
The males are similar to the females. They have the eyes closer together at the top of the head than on the females. The males do not have mandibles in their mouthparts and do not feed on blood.

Distribution
Most tabanid flies have larvae which inhabit areas that are aquatic or very wet. In dry climates tabanids are usually rare, or have a very localized distribution. *Tabanus* occurs in all zoogeographical regions. *Chrysops* occurs in all zoogeographical regions but mainly in the Nearctic, Palaearctic and Oriental Regions. *Haematopota* occurs in the Nearctic, Palaearctic, Afrotropical and Oriental Regions and is present but rare in the Neotropical Region. *Hybomitra* occurs in the Nearctic Region and *Lepiselaga* occurs in the Amazon basin of the Neotropical Region. The greatest variety of tabanid genera are in the Neotropical Region.

Hosts
The species of tabanids which feed on blood have variable feeding preferences. In general *Tabanus* species have a wide host range of mammals, particularly horses and cattle; they also feed on reptiles. *Chrysops* species often have a preference for deer but feed on cattle and other bovids,

camels, humans and monkeys. *Haematopota* species are more specific for species of bovids, including cattle.

Life cycle
This is similar to the *Musca* type (Figure 3.4). Eggs are laid in compact batches on vegetation close to the larval habitat. The larvae burrow into wet soil or similar habitats at the edge of streams or ponds. The larvae are air breathing. Most larvae are carnivorous but some also feed on plant material. Development of the larvae may take from three months to three years. Pupation occurs close to the soil surface and mating takes place soon after emergence of adults from pupae. The females of blood feeding species then seek out hosts for the first blood meal.

Behaviour
Feeding occurs in daylight. Flight is powerful but dispersal is often limited, thus adults of many species remain close to wooded areas or feed at the canopy of forests. Attempts to feed on hosts are persistent despite the defensive behaviour of the host. The bites of tabanids are painful and produce a superficial bloody wound on which the fly feeds.

Disease
HUMANS Species of *Chrysops* transmit the nematode *Loa loa* causing loasis. Tabanids may be significant in the transmission of *Francisella tularensis*, the causative agent of tularaemia. Biting nuisance is often caused by tabanids.

DOMESTIC ANIMALS Species of *Tabanus* transmit the protozoans *Trypanosoma evansi* and *T. vivax* causing trypanosomiasis (of the type known as surra). Species of *Hybomitra* and *Tabanus* transmit the nematode *Elaeophora schneideri* to deer and sometimes sheep causing elaeophorosis. The rickettsia *Anaplasma marginale* is transmitted mechanically by *Tabanus* to cattle. The virus causing equine infectious anaemia can be transmitted by tabanids. A wide variety of other pathogens may be transmitted by various tabanids, including *Francisella tularensis*, the causative organism of tularaemia and *Bacillus anthracis*, the causative organism of anthrax. Substantial biting stress and loss of production can result from tabanid feeding on cattle and sheep. Horses can be greatly stressed by tabanid feeding.

3.3.2 Snipe flies – Rhagionidae

This family of flies is in the Brachycera, the genera *Symphoromyia*, *Atherix*, *Spaniopsis* and *Austroleptis* have species which suck the blood of livestock or humans. They are not involved in transmission of pathogens.

Symphoromyia

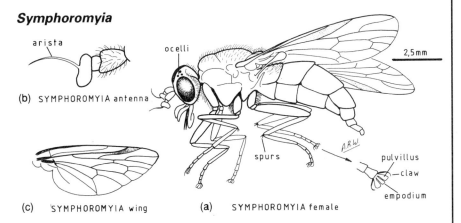

Figure 3.30

- This is a medium sized, elongated, black coloured fly. It has wings and mouthparts similar to tabanid flies.
- The eyes are of medium size and there are three ocelli between the eyes.
- The antennae have three segments, the segment at the far end is large and has an arista as a plain bristle.
- The feet have two pulvilli pads and the central empodium is also in the form of a pad. There are spurs on the tibiae of the hindlegs.
- *Symphoromyia* occurs in Nearctic and Palaearctic Regions, *Atherix* occurs in Nearctic and Neotropical Regions, *Austroleptis* and *Spaniopsis* occur in the Australasian Region.

3.4 CYCLORRHAPHAN FLIES

3.4.1 Introduction to the Cyclorrhaphan flies

Flies that have an antenna with an arista are in the sub-order Cyclorrhapha. An exception is the group of families known as the pupiparan flies (for example Hippoboscidae) which have unusual antennae. There are many species of importance to health in this group. The vast majority of cyclorrhaphan flies are harmless but many are very similar in appearance to those species important to health.

FLIES – DIPTERA

The taxonomy of these flies is complex and they may also be known as Muscomorpha.

Identification of cyclorrhaphan flies important to health is best done with additional evidence of some harmful association with humans or livestock.

The species of importance to health can be grouped as four basic types:

1. The hippoboscid flies (for example *Hippobosca*, Figure 3.31) are specialized as ectoparasites in the adult stage. They have reduced antennae, sometimes no wings, a flattened shape and large claws.
2. The muscid flies (for example *Musca*, Figure 3.33) are characterized by having no bristles on a plate of the thorax known as the hypopleuron (this plate may also be known as the meron). Most genera (but not *Musca*) have vein 4 of the wing in an even curve out to the wing margin (compare Figures 3.33, 3.38, 3.41). The mouthparts may be of the piercing or sponging type. Some species of *Musca*, *Morellia* and *Hippelates* can scrape at skin to feed on blood using the prestomal teeth on the labella of the mouthparts (Figure 3.33). The muscid type includes *Glossina*, *Stomoxys*, *Musca* and *Haematobia*.
3. The calliphorid types (for example *Calliphora*, family Calliphoridae, Figure 3.41) are characterized by a row of bristles on the hypopleuron and vein 4 of the wing having a sharp bend in it, taking the vein to the forward edge of the wing. It is important to distinguish the hypopleural bristles from the mesopleural bristles which are present in muscids as well. Calliphorids have sponging mouthparts. The calliphorids include *Lucilia*, *Phormia*, *Cochliomyia*, *Chrysomya*, *Sarcophaga*, *Wohlfahrtia*, *Auchmeromyia* and *Cordylobia*. The importance to health of these calliphorids is due to their larvae living on or in the tissues of mammals, causing myiasis.
4. The oestrid type (for example *Oestrus*, family Oestridae, Figure 3.50) may have a group of setae on the hypopleuron and vein 4 of the wing is similar to calliphorids except in some species such as *Oestrus* in which it is similar to muscids. The oestrids as adults have mouthparts much reduced and may be unable to feed. They either have a very hairy appearance or pimpled appearance as adults. The oestrids are important to health because of their highly parasitic life as larvae within the tissues of mammals, causing myiasis. The oestrids include *Oestrus*, *Rhinoestrus*, *Cephalopina*, *Gedoelstia* and *Hypoderma*. Other groups causing myiasis are the Cuterebridae (for example *Dermatobia*) and the Gasterophilidae, (*Gasterophilus*).

The identification of flies causing myiasis is usually based on larvae because this is the stage which causes the problem. Thus the figures for the myiasis flies all have larvae most prominent. The larvae shown are mostly the third stage; first and second stage larvae may differ from the third in general shape and be difficult to identify. The most useful feature of larvae for identification is the posterior spiracle. In general, larvae which live in decaying vegetable material and faeces have very curved spiracular slits; those that live in dead or live flesh have straighter slits.

Myiasis is a varied method of parasitism. The basic distinction is between two types: those flies with larvae that can live either on the flesh of dead animals or on the flesh of live animals, causing facultative myiasis; or flies which are highly adapted so that the larvae can only develop in the flesh of live animals, causing obligatory myiasis. There are many species of flies causing myiasis in wild animals and they may occasioally infest domestic animals. Only the species commonly infesting humans and domestic animals are included here.

3.4.2 Louse flies, forest flies, and keds – Hippoboscidae

These flies are often known as hippoboscids. They are not typical cyclorrhaphan flies because of the adaptations of the adults to an ectoparasitic life. The life cyle is also unusual, with larviparous reproduction. The genera *Hippobosca* and *Melophagus* are found on livestock and *Pseudolynchia canariensis* infests domestic pigeons. *Lipoptena* and *Neolipoptena* infest goats and deer. Bats are infested with similar pupiparan flies of the families Nycteribiidae and Streblidae.

Structure

- Hippoboscids can be confusing to identify because the wings may be absent (for example *Melophagus*) or only present whilst the fly is searching for its host (for example *Lipoptena*).
- The antennae are not typical of cyclorrhapan flies (Figure 3.9). They are very compact, in pockets between the eyes, and have a few setae protruding from the outer segment.
- Adult flies have six large stout legs ending in large and sharply curved claws.
- The body is flattened dorso-ventrally but the abdomen may be very stout if it contains a full grown larva.
- The abdomen is distinct from the thorax, but is not clearly segmented.
- The head may appear continuous with the thorax, but close examination will show the division.
- The eyes are small and widely separated.
- The mouthparts are seen as large palps protruding as a close pair in front of or below the head. The piercing mouthparts are retracted within the head when not in use.

FLIES – DIPTERA

Hippobosca

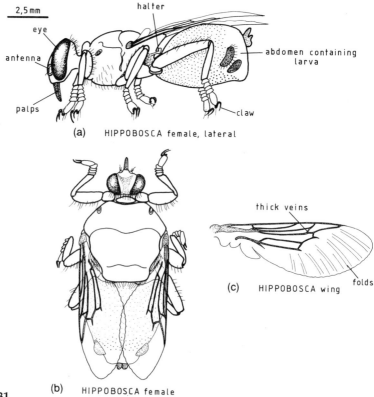

Figure 3.31

(a) HIPPOBOSCA female, lateral
(b) HIPPOBOSCA female
(c) HIPPOBOSCA wing

- These are known as forest flies or louse flies.
- They are large flies with a distinct head and obvious eyes.
- They always have halteres.
- The wings are well developed with very thick veins near the front edge; the rear area of the wing is membranous and with folds like a fan.

Melophagus

- These are known as keds.
- These are medium sized flies with the head not distinct from the thorax.
- There are no wings and no halteres.
- Eyes are inconspicuous.
- The body is covered in bristles and setae.
- *Melophagus ovinus*, the sheep ked, can be mistaken for a tick. However, only larval ticks have six legs and sheep keds are much larger than

CYCLORRHAPHAN FLIES

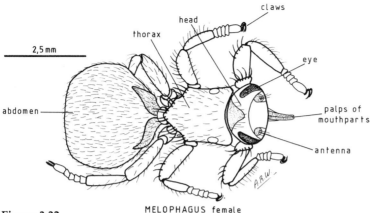

Figure 3.32   MELOPHAGUS female

larval ticks; nymphal and adult ticks have eight legs, and in all ticks the body is not divided into head, thorax and abdomen.

Sexes
These are similar, both are ectoparasitic.

Distribution
Hippobosca species occur in tropical humid, warm humid and dry climates of the Afrotropical, Oriental and Palaearctic Regions. *Melophagus ovinus* occurs in sheep rearing areas in warm humid and cool humid climates throughout the world. It does not occur in hot climates.

Hosts
Hippobosca species infest a wide variety of mammals: *H. rufipes* and *H. variegata* infest cattle and horses, *H. equina* infests horses, *H. camelina* infests camels, *H. longipennis* infests dogs. *Lipoptena capreoli* infests goats. *Melophagus ovinus* infests sheep.

Life cycle
This is the larviparous type, similar to that of *Glossina*, Figure 3.5. Larvae develop singly in the female until fully mature. The larvae are laid on vegetation or in animal housing by *Hippobosca* or on the host by *Melophagus ovinus*. The larvae pupate rapidly. Adults emerge from the puparium and seek out hosts.

Behaviour
Hippobosca species are strong fliers and will fly actively between hosts

within a herd but they do not normally fly long distances. Much time is spent clinging to the hair of the host and taking repeated small blood meals. *Melophagus ovinus* is a permanent ectoparasite of sheep, crawling in the wool and taking repeated small blood meals. Transfer between sheep is by contact.

Disease
HUMANS Hippoboscids flying in search of hosts may land on humans and cause nuisance or possibly attempt to feed. *Melophagus ovinus* may attempt to feed on shepherds.

DOMESTIC ANIMALS *Hippobosca* species cause biting stress to livestock but the importance of this is little known. *Melophagus ovinus* causes loss of production and damage to hides known as cockle. The non-pathogenic protozoan *Trypanosoma melophagium* is transmitted by *M. ovinus*.

3.4.3 House flies – *Musca*

The family Muscidae contains a wide variety of similar genera of importance to humans and domestic animals. These are typical cyclorrhaphan flies.

Identification of most muscid genera and species requires specialist knowledge used in the correct context of geography and hosts.

The representative genus is *Musca*, but other flies of importance to health are: the lesser house fly *Fannia canicularis*; the false stable fly *Muscina stabulans*; also *Atherigona orientalis*; *Hydrotaea* species, and others. The genera *Morellia* and *Hydrotaea* are described below. There are about 60 species of *Musca*, the ones of importance to health are in the *M. domestica* and the *M. sorbens* groups of species.

Structure
- The antenna is of typical cyclorrhaphan type (Figure 3.8), and the arista has setae on both sides.
- A characteristic of muscid flies is the lack of a row of bristles on the plate of the thorax called the hypopleuron (compare Figure 3.33c with calliphorid flies, Figure 3.41). There are bristles on the other plates, mesopleuron and sternopleuron.
- The mouthparts are folded up under the head when not in use (compare Figures 3.33a and 3.33b). Preserved specimens may have mouthparts extended or folded. Extended mouthparts form a thick proboscis, at the end of which are rounded labella lobes with fine grooves in their surface. Liquid food soaks onto these and is then taken into the fly through the labrum.

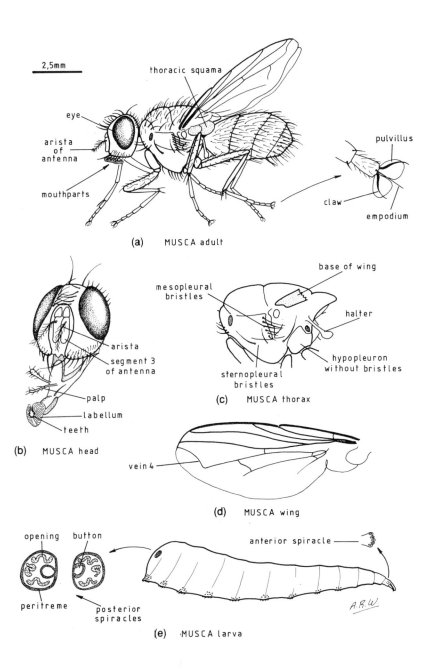

Figure 3.33

- The palps are small.
- The abdomen has four visible segments and other segments concealed at the back.
- The wings have vein 4 sharply curved up to the anterior edge (*Musca* is different from other muscid flies in this characteristic). The squamae on the thorax are large. At rest the wings are held apart over the abdomen.
- The feet have two claws, two pulvilli pads and a central empodium as a fine seta.
- The colouration is dull, not shiny or metallic. There are four dark lines along the top of the thorax and the abdomen has yellow sides.
- The larvae are cylindrical, with clear segmentation, pointed at the anterior end but without a distinct head, Figure 3.33e. The posterior spiracles of the larvae have a complete peritreme with an inner sclerotized ring or button and with sinuous (curvy) openings. The anterior spiracles have five openings.

Sexes
The sexes are similar in appearance, but the eyes of females are further apart than those of males.

Distribution
Some species such as *M. domestica* are very widely distributed throughout the world; this is partly due to their close association with humans. The genus is adapted to a wide variety of climates. Species of the *Musca sorbens* group are closely associated with humans and cattle in the Afrotropical and Oriental Regions. This group includes *M. vetustissima* which occurs in the Australasian Region. *Musca autumnalis*, the face fly, occurs in Afrotropical, Palaearctic and Nearctic Regions. *Musca crassirostris* occurs in the Palaearctic, Afrotropical and Oriental Regions.

Hosts
Most species of *Musca* feed on liquids rich in organic matter to be found on excrement, rubbish, carcasses and on secretions from humans such as from mucous membranes and wounds. The mouthparts of most *Musca* are adapted only for sponging liquid from the surface of a solid. The prestomal teeth on the labella can abrade the surface and may increase the flow of liquids. *Musca crassirostris* has strong prestomal teeth and abrades the skin of cattle to draw blood on which it feeds.

Life cycle
The *Musca* type is shown in Figure 3.4.

Behaviour

Musca flies are active in the day-time. Flies of the *M. domestica* group prefer cool and shaded conditions and are often active inside buildings. Flies of the *M. sorbens* group are active at higher temperatures outside in the sunlight.

Disease

HUMANS The feeding habits and mechanism of feeding makes *Musca* flies (also *Atherigona* and *Hydrotaea*) potential vectors of about 60 pathogens which can be transmitted on contaminated mouthparts. However, *Musca* flies are not usually the most common means by which any of these pathogens are transmitted. Important pathogens include the following:

- viruses of hepatitis and polio;
- bacteria causing salmonellosis, shigellosis, tuberculosis, trachoma, leprosy and yaws;
- the protozoa causing amoebic dysentery;
- the nematodes of threadworm and hookworm.

Musca larvae do not cause true myiasis by penetrating living flesh but if eaten with contaminated food they may survive in the intestine. Invasion of the urinary tract causing urinary myiasis may occur with larvae of *Fannia canicularis*, *Musca domestica* and *M. sorbens*. *Musca sorbens* and *M. vetustissima* cause great nuisance by crawling on the face to feed at the eyes and mouth.

DOMESTIC ANIMALS The situation with transmission of pathogens is similar to pathogens of humans.

- The bacterium *Moraxella bovis* which causes keratoconjunctivitis (pink eye) in cattle is transmitted by *Musca* species.
- Species of the nematode worms *Habronema* which infect horses are transmitted by *Musca*. The nematode worm *Thelazia* which infects the eyes of cattle and horses is transmitted by the face fly *M. autumnalis*, and other species. The nematode *Parafilaria* which infects muscle of cattle is transmitted by *Musca* species.
- *Musca* larvae do not cause true myiasis by penetration of live flesh but they may develop in wool contaminated with grease and faeces.
- Nuisance caused by these flies can be considerable, the face fly *M. autumnalis* is typical. *Musca vetustissima* can scrape at the skin causing superficial wounds and much stress to cattle, sheep and horses.

3.4.4 Sheep head fly (sweat fly) – *Hydrotaea*

This genus is in the family Muscidae, and is an important transmitter of faecal pathogens in tropical climates (synonym: *Ophyra*).

Structure

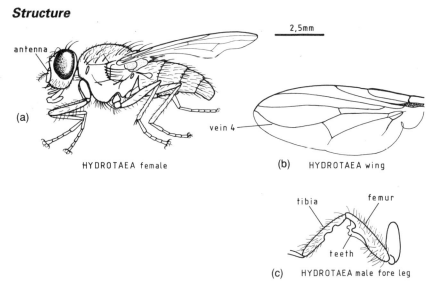

Figure 3.34

- These flies are very similar to *Musca* but vein 4 of the wing makes an even curve out to the tip of the wing (compare Figure 3.34b with 3.33d).
- The antenna is of typical cyclorrhaphan type.
- In males, the femur of the forelegs has teeth which close on corresponding indentations on the tibia.
- The frontal triangle on the head of females usually has a powdery appearance.
- The body colouration is not shiny or metallic, it is dark olive green with yellow near the base of the wings.

Other features

The distribution is worldwide but highest numbers are found in cool humid climates of the Palaearctic and Nearctic Regions. The life cycle is the *Musca* type, with larvae in carrion and decomposing organic material. Adults swarm around the head of the host feeding at wounds (such as fighting wounds on heads of sheep), at the eyes and mouth. The irritation caused is intense. *Hydrotaea* transmits the bacteria (for example *Corynebacterium pyogenes*) which cause mastitis in cattle.

3.4.5 Sweat flies – *Morellia*

This genus is in the family Muscidae.

Structure

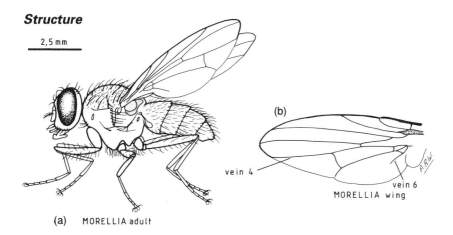

Figure 3.35

- These flies are very similar to *Musca* but vein 4 of the wing makes an even curve out to the tip of the wing. There are two wide dark bands along the top of the thorax.
- The colouration is dark and not metallic.

Other features

Distribution is worldwide. The life cycle is of the *Musca* type. Adults swarm around livestock and sometimes humans, feeding on sweat and at mucous membranes, causing severe nuisance.

3.4.6 Eye flies – *Hippelates* and *Siphunculina*

These genera are in the family Chloropidae of acalyptrate cyclorrhaphan flies. They are separate from the rest of the flies described but are included here for convenience.

Structure

- These are small flies with typical cyclorrhaphan features.
- The antenna is short and with a spherical third segment which has an arista which is either bare or with few short setae.
- The wings are a simpler version of the muscid type, without a vein 6 and with small squamae.
- There is a large spur on the tibia of the hindleg.

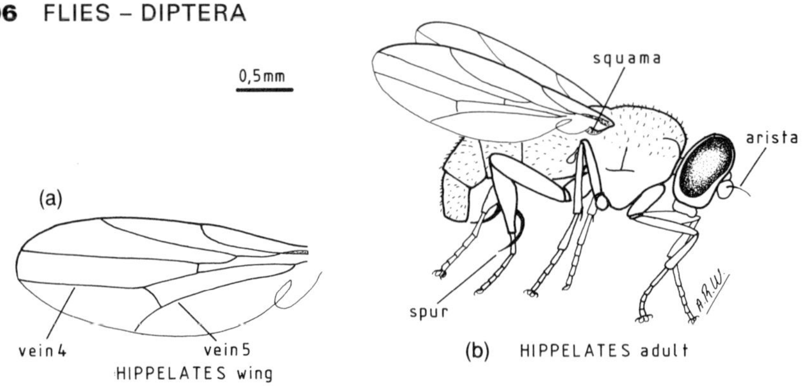

96 FLIES – DIPTERA

Figure 3.36
(a) HIPPELATES wing
(b) HIPPELATES adult

- The triangular plate with the ocelli on the head is large making the eyes far apart.
- There are no bristles on the thorax or abdomen and the colour is dull brown.

Other features
Hippelates occurs in the Neotropical and Nearctic Regions; *Siphunculina* occurs in the Oriental Region. Adult flies swarm around the eyes and head of livestock, and at wounds, feeding on tears, mucous secretions and blood, causing severe nuisance. They may transmit the bacteria causing conjunctivitis and yaws.

3.4.7 Tsetse – *Glossina*

Flies of the genus *Glossina* are known as tsetse. They are closely related to the family Muscidae, but normally given their own family name, Glossinidae. There are 23 species.

Structure
- Antennae are of the cyclorrhaphan type. The arista is very large and with smaller setae on the main setae, which make it look like a feather.
- The long mouthparts protrude in front as a single unit (Figure 3.37a). The mouthparts consist of a pair of thick and long palps and a piercing proboscis. When feeding the proboscis is lowered down from the palps (Figure 3.37b).
- The other common flies with piercing mouthparts protruding forwards are *Stomoxys* and *Haematobia* (Figures 3.38 and 3.39). The palps of *Stomoxys* are much shorter than the proboscis, neither type of fly has its wings folded close together over the abdomen when at rest, and *Haematobia* is much smaller than *Glossina*.

CYCLORRHAPHAN FLIES

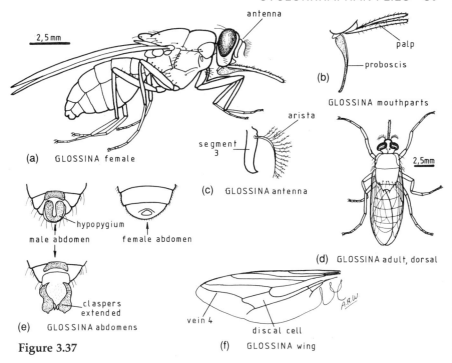

Figure 3.37

- When resting the long wings overlap, covering the abdomen.
- The venation of the wings is similar to muscine and stomoxyine flies (*Musca*, *Stomoxys* etc.), but the discal cell is often known as the hatchet cell because of its shape like a hatchet or axe.
- *Glossina* is similar to muscid flies in having no hypopleural bristles (compare with Figures 3.33 and 3.41).
- Most *Glossina* species are brown coloured.
- The larva can only be seen when fully grown and laid by the female. It burrows immediately into soil and forms an oval shaped puparium (Figure 3.5).

Sexes

Both sexes are active blood feeders. They are similar except for the genital apparatus. On the ventral side of the abdomen the male has a hypopygium like a knob, Figure 3.37e. This can be used to distinguish the three taxonomic and ecological groups of tsetse when stretched out to show the structure of the claspers.

Distribution

Glossina occurs only in the Afrotropical Region. The genus is divided into several ecological and taxonomic groups which are adapted for survival in particular vegetation types. These vary from semi-arid woodland and wooded savanna favoured by the morsitans group, to river banks and wet forest favoured by the palpalis and fusca groups. Usually *Glossina* species are restricted to undisturbed natural vegetation but *G. palpalis* feeding on humans have adapted to farmland around human housing.

Hosts
Females and males feed only on the blood of mammals and reptiles, occasionally on birds. Most species have a preferred host and in addition a fairly wide range of alternative hosts. *Glossina morsitans* prefers species of pig but also feeds on wild ruminants and cattle. *Glossina palpalis* feeds on reptiles but also man. The feeding behaviour of any one species tends to vary in different habitats.

Life cycle
All tsetse are specialized for a larviparous cycle in which one larva at a time develops inside the female (Figure 3.5). Larval growth is supported by repeated blood meals until the larva is of a larger mass than the female. It is then laid on loose soil, into which it burrows. A puparium is formed and an adult develops. A female will produce her first larva in 15 to 20 days and then take 9 to 10 days to produce each subsequent larva.

Behaviour
Tsetse spend most of their time in humid resting sites on branches of trees and bushes. Hosts are found in the day-time by short and fast flights from the resting sites. Tsetse make a loud buzzing noise and attempts to feed are persistent.

Disease
HUMANS The species of *Glossina* that commonly feed on man are important vectors of the protozoans causing human trypanosomiasis or sleeping sickness; other species are also capable of transmission. *Glossina palpalis*, *G. tachinoides*, and *G. fuscipes* transmit *Trypanosoma gambiense* causing Gambian sleeping sickness. *Glossina morsitans*, *G. swynnertoni* and *G. pallidipes* transmit *Trypanosoma rhodesiense* causing Rhodesian sleeping sickness.

DOMESTIC ANIMALS Trypanosomiasis in cattle, sheep, goats, horses, donkeys and camels is mainly caused by the protozoans *Trypanosoma brucei*, *T. congolense* and *T. vivax* which are transmitted by *Glossina morsitans*, *G. pallidipes*, *G. fuscipes*. Other species are also capable of transmission. A trypanosomosis of pigs is also caused by *T. simiae* transmitted mainly by these same three *Glossina* species. These types of trypanosomosis are known as nagana. Although the bites of tsetse are painful the problem of biting stress is usually considered insignificant in comparison with the effects of trypanosomosis.

3.4.8 Stable flies – *Stomoxys*

This genus is within the family Muscidae and has 18 species. A very common species, *Stomoxys calcitrans* is known as the stable fly, or biting house fly. There are about ten genera of flies in a group called the stomoxyine flies (sub-family Stomoxyinae), these include *Stomoxys* and *Haematobia*, and also *Haematobosca* and *Stygeromyia* which feed on the blood of cattle.

Structure

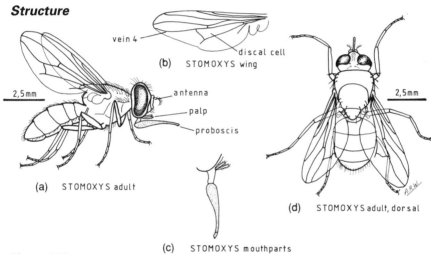

Figure 3.38

- Antennae are of typical cyclorrhaphan type, with sparse setae on the dorsal side of the arista.
- The piercing mouthparts protrude forwards as a single unit when not in use (similar to that of *Glossina*) but the maxillary palps are short and not held together with the proboscis.
- *Stomoxys* is easily mistaken for the common house fly, *Musca domestica*, or the false stable fly *Muscina stabulans*. The distinguishing features of *Stomoxys* are the piercing proboscis protruding forwards and the shorter and broader abdomen. *Stomoxys* species have four dark lines along the top of the thorax like *Musca* but the sides of the abdomen are dark grey or brown.
- When at rest the shining wings are held wide apart and contrast with the dark body. The wings are similar to those of *Glossina* but the discal

- cell has a more rectangular shape and the 4th vein does not curve upwards sharply toward the anterior edge of the wing as in *Musca*.
- There are no bristles on the hypopleuron (compare with Figures 3.33 and 3.41).
- *Stomoxys* species are similar to *Haematobosca* and *Haematobia* but are larger flies having short maxillary palps and different feeding behaviour.
- Larvae are similar to those of *Musca* (Figure 3.33e).

Sexes
The sexes are closely similar and both feed on blood.

Distribution
Stomoxys calcitrans occurs widely in most zoogeographical regions. *Stomoxys niger* and *S. sitiens* are common species in the Afrotropical and Oriental Regions.

Hosts
Stomoxys feeds on a wide variety of hosts including cattle, horses, pigs and humans. Repeated blood-meals are taken during the adult life and often a fly will feed on several hosts in one day.

Life cycle
This is the *Musca* type (Figure 3.4). Eggs are laid scattered over the larval habitats which are wet rotting vegetation, and in the case of *S. calcitrans* in wet dung mixed with vegetation but not in pure dung. Pupation occurs in the larval habitat. Several batches of eggs are produced by each female over a life span of about 20 days.

Behaviour
Adults rest in exposed sunny sites on walls and vegetation, usually close to animal housing. During day-time adults persistently attempt to feed, particularly on the legs of cattle, but fly back to other resting sites between meals. Cattle may be seen to shake one leg after the other to disperse the flies. Very large numbers of these flies may infest cattle.

Disease
HUMANS *Stomoxys* can be a biting pest. They have been implicated in the transmission of numerous pathogens but are not normally the main vector of any pathogen to humans.

DOMESTIC ANIMALS *Stomoxys* can be a severe biting pest of livestock. In cattle this causes loss of gain in liveweight and of milk production. The biting stress to horses and cattle reduces condition and ability to resist disease. *Stomoxys* species have been implicated in the transmission of numerous pathogens but are not normally the main vector. They are known to transmit: the nematode worm *Habronema microstoma* to horses; the protozoan *Trypanosoma evansi* causing trypansomosis (of the type known as surra) in horses, donkeys, camels and cattle; the virus of Potomac horse fever.

3.4.9 Horn fies, buffalo flies and other stomoxyine flies – *Haematobia* and *Haematobosca*

These are other genera of flies closely related to *Stomoxys* which feed on blood of domestic animals. The naming of these genera has had many changes. The usual names of the most important species are: *Haematobia irritans*, the horn fly (also known as *Lyperosia irritans*); *Haematobia exigua*, the horn fly or buffalo fly; *Haematobosca stimulans*. Another related genus with species that feed on blood is *Stygeromyia*.

Structure
- The antennae are of typical cyclorrhaphan type (Figure 3.8).
- These flies are smaller than *Stomoxys*.
- The eyes are relatively large.
- The mouthparts are formed into a piercing proboscis, similar to that of *Stomoxys*, (Figure 3.38), but the palps are larger relative to the proboscis than in *Stomoxys*.
- The wings are typical muscid type but vein 4 does not curve sharply towards the anterior edge as in *Musca*.
- There are no bristles on the hypopleuron (compare with Figures 3.33 and 3.41).
- The larvae are typical muscid type. They develop only in the fresh dung of their hosts.

Haematobia

- Small flies with palps that are as long as the proboscis.
- The antenna has an arista with setae only on the top side.
- *Haematobia* have a resting posture on cattle with wings spread apart to form a triangular shape.

102 FLIES – DIPTERA

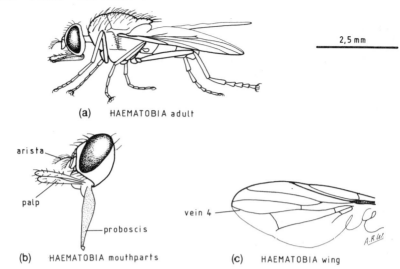

Figure 3.39

Haematobosca

- These are intermediate in size between *Haematobia* and *Stomoxys*.
- The palps are nearly as long as the proboscis and they are shaped like a club.
- The arista has setae on both sides.

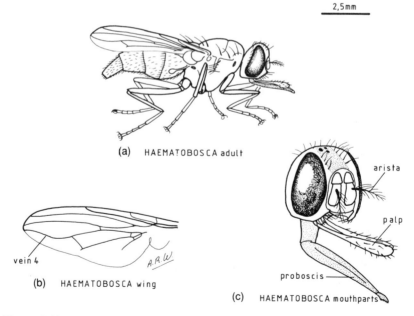

Figure 3.40

Sexes
These are similar in appearance, both feed on blood.

Distribution
Haematobia irritans occurs in warm humid and tropical humid climates of the Palaearctic and Nearctic Regions. *Haematobia exigua* occurs in similar climates but in the Australasian and Oriental Regions. *Haematobosca* occurs in the Afrotropical, Palaearctic and Oriental Regions.

Hosts
Species of both genera feed mainly on the blood of cattle and buffalo. Horses and humans may be used as hosts.

Life cycle
This is the typical muscid type (Figure 3.4). Eggs are laid on fresh dung of cattle or buffalo and mating occurs close to the host.

Behaviour
Haematobia adults spend most time on the backs of cattle and buffalo. They take repeated small meals in any one day and move from place to place on the host by short flights. They can fly strongly to disperse to other hosts. *Haematobosca* species are not so closely associated with their hosts.

Disease
HUMANS *Haematobia* may feed on humans, particularly herders of livestock.

DOMESTIC ANIMALS Massive numbers of flies may infest cattle causing severe biting stress. *Haematobosca* is also found around cattle in sufficient numbers to cause biting stress. The nematode worm *Stephanofilaria stilesi* is transmitted to cattle by *Haematobia irritans* causing hump sore.

3.4.10 Blowflies – *Calliphora, Lucilia, Phormia*

The blowflies are common and conspicuous by their size, buzzing noise and their metallic colouration of thorax and abdomen. The colours vary between species so that it is difficult to identify the genera by colour alone. Blowflies are important in forensic medicine to establish time of death of human corpses. A closely related genus is *Protocalliphora* in which the larvae are blood-sucking on nestling birds within the Nearctic and Palaearctic Regions. *Protophormia terraenovae* is an important blowfly in some areas. Some species of *Lucilia* used to be classified in the genus *Phaenicia*.

General features
- The antenna is of the cyclorrhaphan type (Figure 3.8), with a large third segment hanging down in a groove in the head and a large arista with setae.
- The adult flies are stout, with many large thick bristles on thorax and abdomen.
- The hypopleuron has a row of bristles.
- The wings have strong veins and are clear. Vein 4 of the wings is sharply bent toward the forward edge of the wing (compare this with muscid flies which have a straighter vein 4, except for *Musca* in which vein 4 is similar to calliphorids).
- The mouthparts are of the sponging type, similar to muscid flies. They are folded under the head when not in use. Preserved specimens may show the mouthparts extended or folded.
- The postscutellum is small.
- The larvae have hooked mouthparts and bands of small spines on each segment. The posterior segment has fleshy projections.

Calliphora
- The colouration is often shiny or metallic dark blue, these flies are known as bluebottles.
- They have very large bristles, with a conspicuous group of them on the propleuron.
- The lower (thoracic) squama has fine long setae on its upper surface.
- The stem vein of the wing is without bristles.
- The larvae are very similar to those of *Lucilia* and screwworms but the anterior spiracles have up to 10 openings and on the posterior spiracles the peritreme is thick and with two internal projections.

Lucilia
- The colouration is often shiny or metallic dark green; these flies are known as greenbottles.
- They do not have such large bristles as *Calliphora*, the propleural bristles are not well developed.
- There are no setae on the upper surface of the lower (thoracic) squama.
- The stem vein of the wing is without bristles.
- The larvae have bands of spines on all segments, but these may be

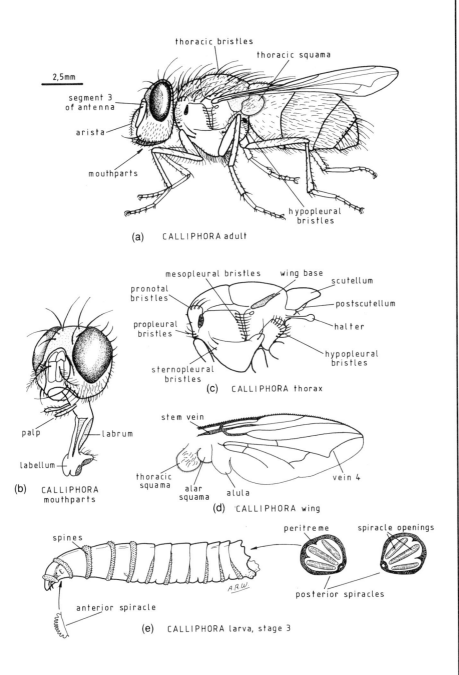

Figure 3.41

106 FLIES – DIPTERA

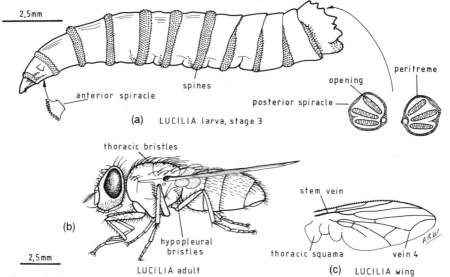

Figure 3.42

obscured at the dorsal surface of anterior segments due to overlap of the segments. The anterior spiracle has 7 to 10 openings. On the posterior spiracle the peritreme is thin and with one internal projection.

Phormia

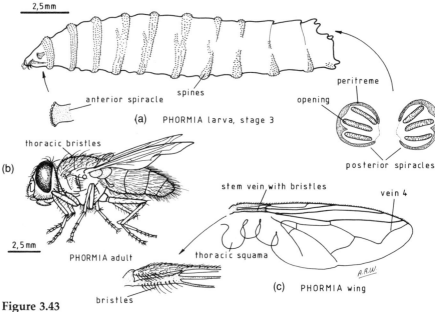

Figure 3.43

- The colour is very dark shiny or metallic blue, often appearing black; these flies are known as blackbottles.
- The anterior spiracle of the thorax is yellow or orange coloured.
- These blowflies are similar in form to *Lucilia* but the long bristles on the thorax are fewer and the short bristles are more numerous.
- There are faint white setae on the upper surface of the lower (thoracic) squama.
- The stem vein of the wing has bristles on the upper surface.
- The larva is similar in shape to *Lucilia* and *Calliphora*; the spines are small, the anterior spiracle has up to 10 openings and the peritreme around the posterior spiracles is incomplete.

Sexes
The sexes are similar in appearance.

Distribution
These genera are most common in the Palaearctic and Nearctic Regions and are the typical blowflies of cool humid climates. They have moved with humans throughout the world and are found in all regions.

Hosts
Blowfly adults feed by sponging liquids, often from the same surface used for egg laying. The eggs are usually laid on decaying animal matter. Larvae usually develop without causing harm to live hosts, but may cause facultative myiasis.

Life cycle
This is similar to that of *Musca* (Figure 3.4).

Behaviour
Blowfly adults are noted for strong flight with a buzzing sound and for gathering at places where there is dead animal material such as slaughterhouses.

Disease
HUMANS These blowflies are household pests and because of their feeding habits have the same potential for disease transmission as *Musca*. However, they do not appear to be as important as *Musca*. Contamination of dirty bandages and wounds may occur causing facultative myiasis. Many blowfly species feed as larvae on a wide variety of carrion including human corpses; this makes them important in forensic medicine (*Manual of Forensic Entomology* by Smith).

DOMESTIC ANIMALS *Calliphora* species (*C. stygia*, *C. augur*, *C. nociva*) are involved in facultative myiasis of sheep known as sheep strike, but in most areas they are less important than *Lucilia* species. *Lucilia* also causes sheep strike. This occurs particularly in woolly breeds where the fleece is easily soiled, thus attracting the adult blowflies. The infestations may be only on the skin surface, but small wounds and soft skin may be invaded and the larvae may spread under the skin. Infestations are of many larvae. The most important species are *Lucilia sericata*, *L. cuprina* and *C. stygia*. *Phormia* species can cause facultative myiasis in sheep, particularly *Phormia regina* and also *Protophormia terraenovae*.

3.4.11 Screwworm flies – *Chrysomya* and *Cochliomyia*

Screwworm flies are very similar to calliphorid blowflies. The larvae of species that are parasitic are called screwworms. This is because the larvae burrow head first into the host, the rings of spines on the larvae making them similar to woodworking screws. There are many species of *Chrysomya* with about 15 occasionally involved in myiasis, but *Chrysomya bezziana* is the predominant cause of myiasis in domestic animals. This species is known as Old World screwworm fly (Old World meaning Palaearctic, Afrotropical, Oriental and Australasian Regions). Similarly there are many species of *Cochliomyia*. Some are obligatory or facultative parasites of minor importance to human or animal health, but *Cochliomyia hominivorax* is predominantly important. This species is known as New World primary screwworm fly (New World meaning Nearctic and Neotropical Regions; 'Primary' because it starts new cases of myiasis). Some species of *Cochliomyia* used to be classified in the genus *Callitroga*.

Structure
- The antenna is of the cyclorrhaphan type (Figure 3.8) with a large third segment hanging down in a groove in the head and a large arista with setae on both sides.
- The adult flies are stout, with some large thick bristles on thorax and abdomen, but these are fewer than in other blowflies.
- The hypopleuron has bristles.
- The wings are usually clear. They have strong veins and vein 4 is sharply bent forwards.
- The mouthparts are of the sponging type, folded under the head when not in use.
- The larvae have hooked mouthparts and bands of small spines at each segment. There may be fleshy projections on most segments or only on the last segment.

Identification of species of screwworm should be confirmed by a specialist taxonomist but details of two species are given here because of the importance of detecting invasions to new areas.

Chrysomya

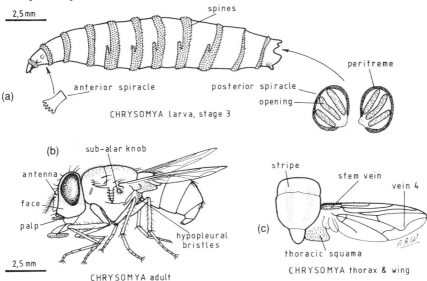

Figure 3.44

- Larvae may have fleshy processes along the body or only on the last segment.
- The projections on the anterior spiracles of larvae vary from 4 to 13 and on the posterior spiracles the peritreme is incomplete and the button is indistinct.
- The bands of spines of larvae are complete on all segments but narrow dorsally on segment 11.
- The adult flies usually have a blue or green colour.
- The wing has the stem vein with bristles.
- The lower or thoracic squama is a large triangular shape, coloured white and with setae on the entire top surface.

Chrysomya bezziana
- The adults of *Chrysomya bezziana* have a pale coloured face. The thorax has two faint dark stripes along the top. The general body colour is blue or blue green.
- The anterior spiracle of the thorax of adults is dark coloured.
- The thorax has a sub-alar knob with setae on it.

FLIES – DIPTERA

- The palps of the mouthparts are small but visible.
- The larvae have four to six projections on the anterior spiracle and fleshy projections on the last segment only, the openings of the posterior spiracles are almost parallel to each other.

Cochliomyia

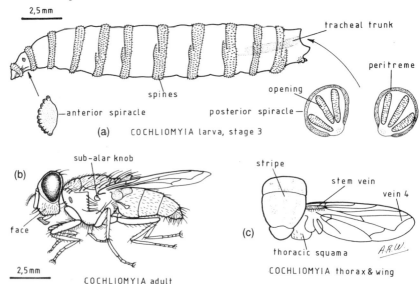

Figure 3.45

- The larvae have 7 to 10 projections on the anterior spiracle and on the posterior spiracle the peritreme is incomplete and the button indistinct.
- Bands of spines on the larvae are on all segments but may be incomplete dorsally on segment 11.
- The adult flies are usually of blue or green colour.
- The stem vein has a row of bristles.
- The thoracic squama has few setae, mostly at the base.

Cochliomyia hominivorax
- The larvae have anterior spiracles with 10 openings and the openings of the posterior spiracles are at an angle to each other.
- Larvae can be distinguished from the those of similar species by the dark colour of the tracheal trunks. These are the air tubes which extend forward in the body from the posterior spiracles. They are pigmented along the length from the last segment to about the ninth segment. This is best seen in fresh specimens.
- The larvae of *Cochliomyia macellaria*, the secondary screwworm, often

infests wounds made by *C. hominivorax*; their mature larvae are very similar, but *C. macellaria* are without pigmented tracheae.
- The adults of *Cochliomyia hominivorax* have a yellow coloured face. They have three conspicuous dark stripes along the top of the thorax, of which the centre one is shorter in the anterior direction. The abdomen is without white spots on the last segment, but *C. macellaria* has such spots.
- The palps of the mouthparts are small and hidden in the groove which contains the mouthparts.
- The sub-alar knob is without setae.
- The thoracic squama has few hairs, the stem vein of the wing has a row of hairs.

Sexes
The sexes of adult flies are similar but the males have eyes close together at the top of the head.

Distribution
Typically *Chrysomya* is confined to the Afrotropical, Oriental, Palaearctic and Australasian Regions. Within these regions they are most common in tropical humid climates. Typically *Cochliomyia* is confined to the Nearctic and Neotropical Regions where they are most common in tropical humid climates. In 1988 *Cochliomyia hominivorax* spread across the Atlantic Ocean to Libya in a shipment of sheep. The outbreak was eradicated, but the potential for screwworm flies to spread is considered a great problem.

Hosts
The obligate parasitic species have larvae feeding on a wide range of mammals including livestock and dogs.

Life cycle
This is basically of the *Musca* type (Figure 3.4). The adults have sponging mouthparts. The females use these to obtain the protein to support egg production from wounds or decaying animal matter. Eggs are laid at the adult feeding site. The obligate parasitic species burrow into the wound area but usually remain together and with their posterior spiracles at the surface. When mature the larvae leave the wound and burrow in loose soil to pupate.

Behaviour
The adults are attracted to the smell of wounds or decaying animal matter but they do not trouble the hosts. Adults are strong fliers and can disperse long distances. They are only active in summer or in warm climates.

Disease
HUMANS *Chrysomya bezziana* is a cause of myiasis in humans particularly in the Oriental Region. *Cochliomyia hominivorax* can cause myiasis in humans. Where the infestation is in the head, starting at a wound or discharge from the nose, it is likely to be fatal. Infestations take the form of a batch of larvae forming a suppurating superficial lesion in muscle or the body wall.

DOMESTIC ANIMALS *Chrysomya bezziana* causes myiasis in cattle and other livestock, the results are serious and often fatal but the frequency of such myiasis varies greatly from area to area. *Cochliomyia hominivorax* is a common cause of myiasis in cattle and other livestock; the results are serious and often fatal. Screwworm infestations on livestock are usually in the form of a large rounded lesion open to the surface where 200 to 300 larvae from one egg batch remain together feeding on the body wall and superficial muscles. The nasal passage may also be infested.

3.4.12 Fleshflies – *Sarcophaga* and *Wohlfahrtia*

These genera are a group within the Calliphoridae family or alternatively they are given the status of the family Sarcophagidae. Unlike other blowflies they do not have shiny metallic coloured abdomens. The abdomens have dull patterns of grey and black. Most species feed as larvae on carcasses, debris or faeces. There are many species of *Sarcophaga* and the taxonomy of this genus is complex. *Sarcophaga cruentata* (= *S. haemorrhoidalis*) larvae are feeders of the facultative myiasis type. *Wohlfahrtia vigil*, *W. nuba* and *W. magnifica* larvae are feeders of the obligate myiasis type.

Structure

- Both genera have adults which are large and bristly and with long legs.
- In contrast to blowflies (Figures 3.41 to 3.43) the body is elongated and has a grey powdery covering with patterns of stripes on the thorax, and of patches on the abdomen.
- The wings are of the calliphorid type and hypopleural bristles are present on the thorax.

- The antenna is of the cylorrhaphan type (Figure 3.8).
- The stem vein is without bristles.

Sarcophaga

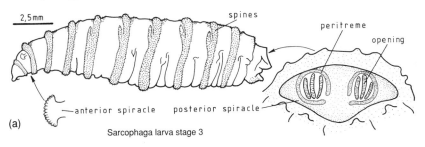

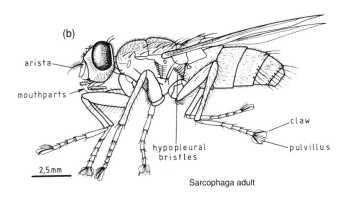

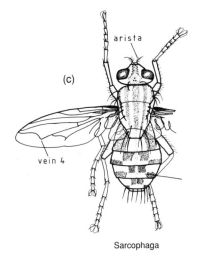

Figure 3.46

114 FLIES – DIPTERA

- Adults have three dark stripes along the top of the thorax and squared or chequered patterns of dark and grey patches on the top of the abdomen.
- On the antenna the arista is hairy at the base but plain at the far end.
- The postscutellum of the thorax is small and the pulvilli and claws of the feet are large.
- The last abdominal segment of *S. cruentata* (= *S. haemorrhoidalis*) is coloured red brown.
- The larva is similar to those of blowflies, with bands of small spines on 12 segments. The anterior spiracles have numerous openings and posterior spiracles are within a deep pocket on the last segment. The posterior spiracle plate has an incomplete peritreme and three straight and vertical openings.

Wohlfahrtia

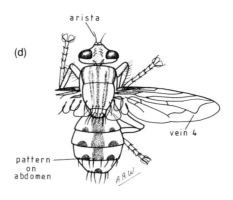

Wohlfahrtia

Figure 3.47

- Adults resemble *Sarcophaga* but the pattern on the abdomen is of separate rounded dark patches.
- On the antenna the arista is without setae.
- *Wohlfahrtia* larvae closely resemble those of *Sarcophaga*. One way to differentiate them is to rear mature larvae to adults.

Sexes
The sexes are similar.

Distribution
Sarcophaga cruentata (= *S. haemorrhoidalis*) is associated with humans and has become widely distributed in the world. *Wohlfahrtia magnifica* occurs

in humid and tropical humid climates of the Palaearctic and Oriental Regions. *Wohlfahrtia nuba* occurs in dry climates of the Palaearctic and Oriental Regions. *Wohlfahrtia vigil* occurs in the Nearctic Region.

Hosts
Larvae of *Sarcophaga cruentata* (= *S. haemorrhoidalis*) are associated with humans. Larvae of *Wohlfahrtia* infest humans and many species of livestock: cattle, horses and donkeys, sheep and goats, camels, pigs, dogs and cats, rodents and birds.

Life cycle
Sarcophaga and *Wohlfahrtia* have a life cycle similar to the *Musca* type (Figure 3.4) but the larvae are produced directly by the female or hatch from the egg immediately after laying. *Sarcophaga* larvae are deposited on dead organic material or on wounds of a live host. *Wohlfahrtia* eggs or larvae are deposited on small wounds or mucous membranes of the host and these larvae can penetrate intact skin.

Behaviour
Sarcophaga females will seek foul smelling material for depositing larvae. *Wohlfahrtia* females will seek out live hosts, not necessarily with the factor of foul smelling wounds.

Disease
HUMANS *Sarcophaga cruentata* (= *S. haemorrhoidalis*) is sometimes a parasite of humans, superficially infesting existing wounds or occurring in the intestine by accidental ingestion. *Wohlfahrtia magnifica* may infest humans, causing superficial myiasis, particularly at the nose or eyes. Infestations are usually of few larvae, but the results can be very serious.

DOMESTIC ANIMALS *Wohlfahrtia magnifica* and *W. nuba* larvae cause superficial myiasis. Usually there are few larvae at an infestation but they form a large wound with serious consequences to the host. *Wohlfahrtia vigil* larvae cause myiasis similar to that of the abscesses formed by *Dermatobia*. Individual larvae occur under the skin with an opening where the posterior spiracles are exposed.

3.4.13 Tumbu fly and floor maggot – *Cordylobia* and *Auchmeromyia*

These genera are calliphorid flies with several species being parasitic on humans and other animals. *Cordylobia* larvae are typical myiasis type.

Auchmeromyia larvae feed on blood but do not remain in the host. This type of parasitism may be classified as myiasis. *Auchmeromyia senegalensis* feeds on humans; it may alternatively be known as *A. luteola*. The adults are difficult to distinguish from many other types of calliphorid flies.

There are two important species of *Cordylobia*, *C. anthropophaga* and *C. rodhaini* (formerly *Stasisia rodhaini*). Other calliphorid flies similar to *Cordylobia* and causing myiasis are *Booponus intonsus* which infests the feet of water buffalo, cattle and goats in the Philippines and Sulawesi; also *Elephantoloemus indicus* which infests the skin of Indian elephants.

Structure

- The antenna is of cyclorrhaphan type (Figure 3.8) and the arista has setae on both sides.
- The adults of both genera are large but they do not have the shiny metallic colour of the blowflies, the colour is dull yellow brown or red brown.
- The wings are of calliphorid type and the thoracic squamae are without setae.
- The stem vein of the wing is without bristles.
- The thorax has distinct hypopleural bristles, the bristles on the thorax are sparse and there are fine setae between the bristles.
- There are four visible abdominal segments.
- The mouthparts are fully developed.

Cordylobia

- In adults the second segment of the abdomen is of similar size to the first and third segments.
- The larvae are stout and covered with small sharp spines.
- The anterior spiracles of the larvae are in the form of a clump of openings; the posterior spiracles are superficial and have curved openings and the peritreme is indistinct. The openings of the posterior spiracles of *C. anthropophaga* are slightly curved, those of *C. rodhaini* are in complex curves.

Auchmeromyia senegalensis (A. luteola)

- In female flies the second segment of the abdomen is much longer than the first or third segments.
- Larvae are dark coloured.
- Larvae have no spines but on the ventral surface there are rough tubercles or bumps on the central segments.

CYCLORRHAPHAN FLIES 117

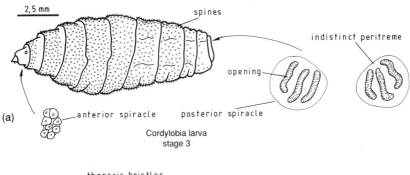

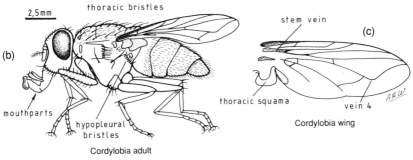

Figure 3.48

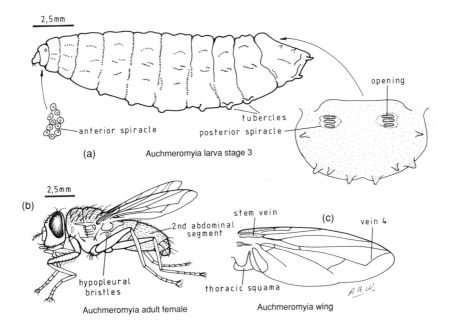

Figure 3.49

118 FLIES – DIPTERA

- Larval segments are divided by rows of spots.
- The anterior spiracles of larvae are in the form of a clump of openings. The posterior spiracles have three straight and horizontal openings, they are superficial and far apart, and the peritremes are indistinct.

Sexes
The sexes are similar except that in males of *A. senegalensis* the second abdominal segment is smaller than that in the female.

Distribution
Both genera occur only in the Afrotropical Region where they are widespread. The adults lay eggs on soil near or within the housing or burrows of their hosts.

Hosts
Larvae of *Cordylobia anthropophaga* infest humans, dogs, cats and a wide variety of wild mammals. *Cordylobia rodhaini* infests antelopes and rodents and infestations of humans have been recorded. Larvae of *Auchmeromyia* feed on pigs and warthogs, and *A. senegalensis* has also adapted for feeding mainly on humans.

Life cycle
This is of the *Musca* type (Figure 3.4), but in contrast to other flies causing myiasis, the eggs are laid on soil. The larvae of *Cordylobia* then seek and penetrate a host. The larvae of *Auchmeromyia* remain free-living and take repeated blood-meals from their hosts.

Behaviour
Females of both genera lay eggs on soil or in animal burrows and are not likely to come into contact with the larval hosts. However females of *C. anthropophaga* may be found laying eggs on soiled clothing, or washed clothing hung to dry in shady places. The adults feed on plants and on dead animal material and excreta. The larvae of *Cordylobia* can remain actively seeking a host for days, and larvae of *Auchmeromyia* spend most of their time off the host.

Disease
HUMANS Larvae of *Cordylobia anthropophaga* penetrate the skin and develop in the subcutaneous tissue as separate swellings like large abscesses. The posterior spiracles of the larvae are exposed through holes in the skin. This is known as furuncular myiasis. The infestations

are usually on the back, buttocks or limbs. They are very painful; fever and malaise also occur. *Cordylobia rodhaini* infestations of humans are rare but cause larger and more serious lesions than *C. anthropophaga*. *Auchmeromyia senegalensis* larvae take repeated small blood-meals from people sleeping on the ground. They cause irritation but no pathogens are known to be transmitted.

DOMESTIC ANIMALS Dogs and cats become infested with *C. anthropophaga* causing disease similar to that in humans. *Auchmeromyia* larvae may be involved in mechanical transmission of trypanosome protozoa.

3.4.14 Nasal bot flies – *Oestrus, Rhinoestrus* and *Cephalopina*

The family of myiasis flies known as Oestridae or oestrids or nasal bot flies has many genera and species. The larvae of all species are obligatory parasites, usually of ungulate mammals but occasionally of humans. The larvae burrow into the nasal cavity or throat. Warble flies are also in the Oestridae. The genera of main veterinary importance are *Oestrus* infesting sheep, goats and antelope; *Rhinoestrus* infesting horses, donkeys and antelopes; and *Cephalopina* infesting camels (*Cephalopina* used to be known as *Cephalopsis*). Other genera of nasal bots include *Gedoelstia* infesting antelope and sheep; *Przhevalskiana* infesting antelope and goats; *Cephenemyia* infesting deer; and *Pharyngobolus* infesting the throat of African elephants.

Structure
- Adults have a large broad head with small widely separated eyes. The eyes of males are closer together than in females.
- The antennae are of the cyclorrhaphan type (Figure 3.8) but the segments are small and the arista is large and without setae.
- The hypopleuron has setae but these are not obvious as hypopleural bristles.
- The legs are small in proportion to the body.
- The thoracic squamae are large.
- Adult mouthparts are reduced to small knobs.
- The larvae have large mouth hooks, bands of spines on most segments and spiracles in which the openings are many small holes.
- Anterior spiracles are not visible on larvae.

120 FLIES – DIPTERA

Oestrus

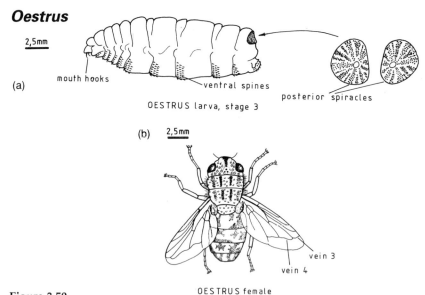

Figure 3.50

- The adults have a head and thorax with pits containing small bumps or tubercles. These tubercles have setae but the body does not have a very hairy or bristly appearance.
- The postscutellum is large.
- The apical cell of the wing is closed by vein 4 joining vein 3 before the wing margin.
- Larvae have bands of spines on the ventral side of each segment, large hooked mouthparts, no fleshy projections, and the spiracular plate is rounded without an indentation.

Rhinoestrus

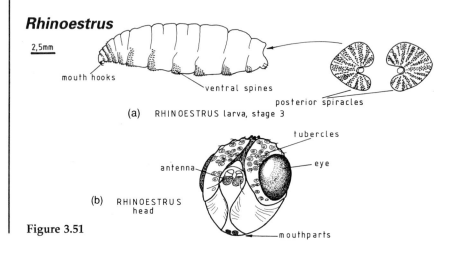

Figure 3.51

- The adults are similar to *Oestrus* but the head and thorax have more conspicuous bumps or tubercles.
- The larvae have bands of spines on the ventral surface of each segment. These bands extend to the lateral and dorsal surface of anterior segments.
- The posterior spiracular plate of the larva has a deep indentation.

Cephalopina

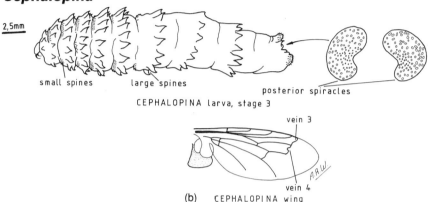

Figure 3.52

- The adults are similar to those of *Oestrus* but the tubercles on the head and thorax are not so conspicuous.
- Vein 4 of the wing has a distinct double bend as it joins vein 3.
- The larvae have bands of large projections on each segment in the form of very large spines. They also have bands of very small spines on both dorsal and ventral surfaces of anterior segments.
- The posterior spiracle of the larva has openings irregularly spaced and a shallow indentation in the spiracular plate.

Sexes
The sexes of adult flies are similar, in males the eyes are closer together than in females.

Distribution
Oestrus is widely distributed in most regions and climates. *Rhinoestrus* is distributed in the Palaearctic, Oriental and Afrotropical Regions. *Cephalopina titillator* is found in areas with large numbers of camels, typically in dry climates of the Palaearctic and Oriental Regions.

Hosts

Oestrus ovis mainly infests sheep and goats; *Rhinoestrus purpureus* infests horses and donkeys; *Cephalopina titillator* infests camels (bactrian and dromedaries).

Life cycle

Oestrus ovis females develop first stage larvae in their bodies (larviparous reproduction). They then flick small batches of larvae into the nose of their host. Larvae of nasal bot flies remain within the nasal tissues of their hosts, also sometimes in the throat tissues. The mature larvae are sneezed out onto the ground where they burrow under the surface then pupate. The larvae contain sufficient food reserves to support the adults. The adults cannot feed.

Behaviour

The adult flies accumulate near their hosts and cause irritation and avoidance behaviour of the hosts.

Disease

HUMANS Infestation by nasal flies is rare and not part of the usual life cycle. Larvae of *Oestrus*, *Rhinoestrus* and *Gedoelstia* may infest people living in close association with sheep or goats. The infestation is typically in the eye or eye socket (ophthalmomyiasis) where the results can be very serious if the larvae are not quickly extracted.

DOMESTIC ANIMALS Infestations of nasal bots are usually of one or a few larvae per host. They may occur in larger numbers and *Cephalopina titillator* has been found as infestations of more than 100 larvae in a single camel. The larvae burrow in the mucous membranes of the nose and throat. Infestations of one or several larvae do not cause severe problems but larger numbers can cause physical damage, stress, loss of production or death. *Gedoelstia* species may infest the eyes of sheep, causing lumpy eye.

3.4.15 Warbles and stomach bots – *Hypoderma* and *Gasterophilus*

These genera are highly specialized for obligatory parasitism. They consist of few species. *Hypoderma* warble flies are classified in the Oestridae family of bot flies; closely related is *Oedemagena tarandi*, the warble fly of reindeer. *Gasterophilus* stomach bot flies of horses are sometimes classified with the oestrids but are usually placed in the family

Gasterophilidae. Other stomach bots of the Gasterophilidae are *Gyrostigma* species infesting rhinoceros, *Platycobboldia loxodontis* infesting African elephants and *Cobboldia elephantis* infesting Indian elephants.

Hypoderma

Structure

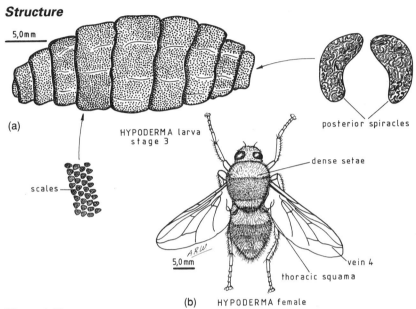

Figure 3.53

- The adults are large stout flies. They have a dense covering of fine setae on the body; these setae have bands of yellow colour and give the flies the appearance of bees.
- The mouthparts are small and without palps.
- The eyes are small, and widely separated in the females.
- The antennae are of cyclorrhaphan type (Figure 3.8); they are small but the second segment is larger than in other oestrid flies and partially covers the third segment.
- The postscutellum is large and hypopleural bristles are a group of hairs.
- The thoracic squama is large and vein 4 of the wing extends to the anterior margin of the wing.
- Mature larvae of *Hypoderma* are large and very stout. The integument is covered in thick flat spines or scales. The characters of such spines can be used to identify species. First stage larvae are proportionately thinner, translucent and without the thick scaled skin of the third stage larvae.

124 FLIES – DIPTERA

- The mouthhooks of larvae are very small, mature larvae have no mouthhooks.
- Posterior spiracles of larvae are indented and with the openings as numerous small curved slits.

Gasterophilus

Structure

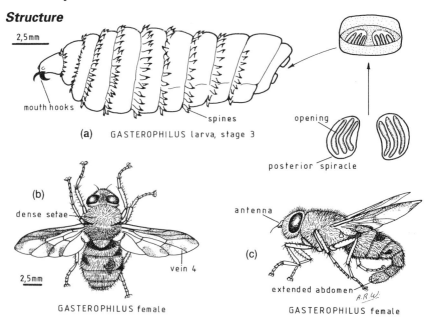

(a) GASTEROPHILUS larva, stage 3

posterior spiracle

(b) GASTEROPHILUS female

(c) GASTEROPHILUS female

Figure 3.54

- The adults are large stout flies with a dense covering of setae.
- The females have the posterior abdominal segments extended as an ovipositor; this, in combination with the dense setae may lead to them being confused with bees.
- The eyes are small, and widely separated in females.
- The antennae are of cyclorrhaphan type (Figure 3.8) but small.
- The thoracic squama is small. Vein 4 of the wing curves down to join the posterior margin of the wing, and the wing has dark patterned areas.
- The mouthparts are very small.
- Mature larvae are large and stout and there are bands of large spines on all segments except the last. The characteristics of such spines can be used to identify species. First stage larvae are proportionately thinner and with bristles instead of spines.
- The mouthhooks are large.

- Posterior spiracles are enclosed in a pocket, and the spiracle openings are curved slits.

Sexes
The sexes of *Hypoderma* are similar. In females of *Gasterophilus* the last abdominal segments are extended as an ovipositor.

Distribution
Hypoderma occurs in the Palaearctic and Nearctic Regions, in warm humid and cool humid climates. *Gasterophilus* originated in the Afrotropical Region but has spread to most other regions with horses, and ranges from tropical humid to cool humid climates.

Hosts
Hypoderma bovis and *H. lineatum* infest cattle; *H. lineatum* may infest horses but the life cycle is not completed. *Hypoderma diana* infests deer; it may infest sheep but the life cyle is not completed. *Gasterophilus* infests horses, donkeys, mules and zebras. Humans may become infested by both genera but do not support full development of the larvae.

Life cycle
Eggs are laid on hairs of the host and the larvae are entirely parasitic. The adults live for a few days whilst mating and laying eggs. Adults do not feed and are supported by reserves from the larval stage.

Hypoderma larvae burrow through intact skin and migrate through connective tissue, for example along the oesophagus and the spine. Finally, mature larvae stop beneath the skin along the back. The resulting lumps in the skin are pierced with a small breathing and exit hole. This appearance of the skin is characteristic of warbles. Infestation of this type is known as furuncular myiasis. The larvae emerge, drop to the ground and pupate.

Gasterophilus eggs or larvae become parasitic when licked by the host, thus the infestation starts in the mouth or throat. The larvae eventually migrate to the stomach where they attach to the inner lining with their bodies exposed to the stomach cavity. When mature they exit in the faeces, then pupate. Larvae of *G. haemorrhoidalis* attach to the rectal wall.

Behaviour
Adult females of some species of both genera are recognized by their hosts as they persistently fly around the host attempting to lay eggs. The hosts may take extreme avoiding reactions and panic (sometimes called gadding).

Disease

HUMANS Infestations by *Hypoderma* and *Gasterophilus* are rare and usually in people working with the natural hosts. The larvae do not develop beyond the first stage in humans. They usually remain superficial, forming a red track beneath the skin, known as a creeping eruption. The larvae may occasionally penetrate deeper, causing much distress and may infest the eye (ophthalmomyiasis) where they are malign and can destroy the eye if not extracted rapidly.

DOMESTIC ANIMALS Species of both genera may cause damage to their hosts when they panic to avoid the egg laying females. *Hypoderma* larvae may damage the muscle and hides sufficiently for the carcass and hide to be downgraded commercially. The stress of infestation can cause poor gain in weight. *Gasterophilus* infestations rarely cause problems but heavy infestation of the stomach may cause secondary complications.

3.4.16 The Torsalo and rodent bots – *Dermatobia* and *Cuterebra*

These genera are in the family Cuterebridae. They are highly specialized obligate parasites. There are few species. *Dermatobia hominis* is also known as the human bot fly, berne or nuche. Despite the human associations of *D. hominis* it is most important as a common and serious infestation of cattle. *Cuterebra* is rarer as a medical or veterinary problem, being specialized for myiasis on rodents or rabbits, but it may infest humans and small livestock. *Neocuterebra squamosa* infests the feet of African elephants.

Dermatobia

Structure

- Adults are large with a metallic blue abdomen and yellow brown thorax and head.
- There are no bristles, but sparse short setae cover the thorax.
- The head projects forwards and has long sides (face), and the eyes are small, and widely separated in females.
- The arista of the antenna has setae on the outer side only.
- The mouthparts are very small and without palps.
- The postscutellum is small.
- The wings are of the calliphorid type with vein 4 joining the anterior margin of the wing close to vein 3. The thoracic squamae are large and without setae.
- The larvae vary depending on the stage.

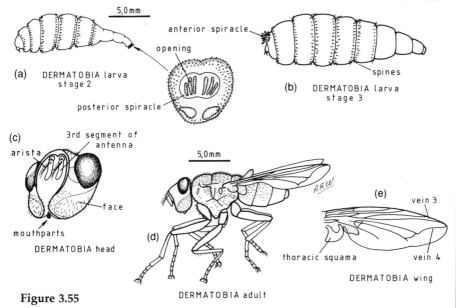

Figure 3.55

- Second stage larvae have stout and clearly defined anterior segments with large but sparse spines. The posterior segments are narrow, poorly defined and without spines. The posterior spiracles have three openings each and are without an obvious peritreme.
- Third stage larvae are stout throughout, with spines on the anterior segments and medium sized mouthhooks. Anterior spiracles are present as two groups of protruding branches but these may not be obvious in preserved specimens.

Cuterebra

Structure

- Adults are large and dark coloured. There are no bristles and the thorax and abdomen is densely covered in short setae.
- The head projects forwards but not as much as in *Dermatobia*.
- Eyes are fairly small and widely separated in the females.
- The antenna has an arista with setae only on the outer surface.
- Mouthparts are very small and without palps.
- The postscutellum is small.
- Wings are of the calliphorid type, with a uniform dark colour, and have a very large alula.
- The mature larvae are stout with a dense covering of spines in the form of scales with serrated edges.
- The larval mouthhooks are of medium size and the posterior spiracles

128 FLIES – DIPTERA

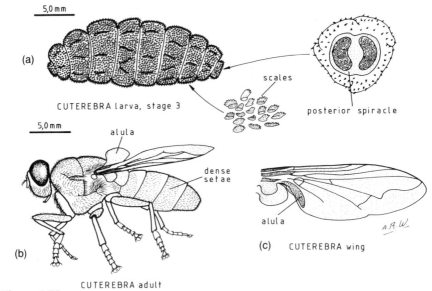

Figure 3.56

are without an obvious peritreme and have many openings as very curved slits.
■ These larvae are similar to those of *Hypoderma* but the scales are different.

Sexes
These are similar except for more widely separated eyes in females.

Distribution
Dermatobia hominis occurs in the tropical humid and warm humid climates of the Neotropical Region; infestations of larvae may be carried to other regions by the hosts. *Cuterebra* occurs in the warm humid and dry climates of the Nearctic Region.

Hosts
Dermatobia hominis larvae infest cattle, sheep, goats, pigs, monkeys, birds, cats, dogs and humans. *Cuterebra* larvae usually infest rodents and rabbits but may infest cats and dogs and infestations of humans have been recorded.

Life cycle
Both genera have life cycles basically of the *Musca* type (Figure 3.4). *Dermatobia hominis* has an additional specialization. The female fly seeks

other insects which feed on cattle and other hosts. The eggs are laid on the other insect and the eggs hatch immediately on contact with the host. The larvae penetrate the skin and develop. Such movement of one animal on another is called phoresy. Many different types of insects are used, some examples are *Psorophora* mosquitoes, *Musca* and *Stomoxys* flies. Some accounts of *D. hominis* describe four larval stages.

Behaviour
The adults of *Dermatobia* and *Cuterebra* do not feed.

Disease
HUMANS Infestations with larvae of *D. hominis* are rare but this species can complete its development in humans. Infestation with *Cuterebra* is very rare. The larvae form and remain in individual swellings like a large abscess in the subcutaneous tissue.

DOMESTIC ANIMALS Larvae of *Dermatobia* and *Cuterebra* develop individually in subcutaneous tissue, remaining at one site until development is complete. The swellings are like large abscesses (furuncular myiasis) and the posterior of the larvae may be exposed through a hole in the skin. Cattle may become infested with many *D. hominis* larvae resulting in stress, loss of production, hide damage or even death.

4 LICE – PHTHIRAPTERA

4.1 INTRODUCTION TO THE LICE

Lice are highly characteristic ectoparasites of mammals and birds. They are recognizable as insects having: a segmented body divided into head, thorax and abdomen; three pairs of jointed legs; a pair of antennae; never any wings. The whole body is flattened dorso-ventrally. The life cycle has an incomplete metamorphosis. The female lays eggs on the host, usually cementing the eggs onto hairs or feathers (Figure 4.3b). The stage which hatches from the egg is the first nymph. This resembles the adults and feeds with them. After growth the nymphs moult through three stages and the final moult produces the sexually mature adults. The adults appear similar except for the external genital structures (Figure 4.1b and c). The whole life cycle is spent on the host and spread between hosts is accomplished by the lice crawling when the hosts are in contact.

The number of genera and species of lice is large. Their classification is

Table 4.1 Summary of genera of lice found in humans and domestic animals

Host organism	Lice genera
Humans	*Pediculus, Pthirus*
Cattle	*Linognathus, Haematopinus, Solenopotes, Damalinia*
Sheep	*Linognathus, Damalinia*
Goats	*Linognathus, Damalinia*
Pigs	*Haematopinus*
Horses and donkeys	*Haematopinus, Damalainia*
Dogs	*Linognathus, Trichodectes, Heterodoxus*
Cats	*Felicola*
Chickens	*Goniocotes, Goniodes, Lipeurus, Cuclotogaster, Menacanthus, Menopon*
Turkeys	*Oxylipeurus, Chelopistes*

complex and uncertain and the features used to distinguish some genera are difficult to describe. The classification used here divides the order Pthiraptera into one sub-order of sucking lice, Anoplura; and the two sub-orders of chewing lice, Ischnocera and Amblycera. Another classification uses Anoplura for the whole order, divided into Siphunculata (sucking lice) and Mallophaga (chewing lice). Names of genera tend to vary depending on the source of information and the host, for example *Damalinia, Bovicola* and *Trichodectes*.

> *Lice are usually only found on their preferred host, thus information on the host and the site of infestation of the lice on that host is an important aid to identification.*

4.2 SUCKING LICE – ANOPLURA

4.2.1 General features

- The Anoplura have mouthparts that are highly adapted for piercing the skin of their host, with a fine set of cutting and sucking structures. These lice take frequent small blood-meals. The mouthparts are usually withdrawn into the head so that all that can be seen of them is their outline in the head or their tip protruding.
- The mouthparts have no palps.
- The thoracic segments are joined together in one unit.
- The abdomen may have sclerotized paratergal plates along the sides. Spiracles are present at the sides of some or most of the abdominal and thoracic segments.
- The tarsus has one segment which bears a single claw, the claw closes onto a projection of the tibia called the tibial spur. The tarsus and tibia may be separated by a visible joint or the two segments may appear as one unit. This structure of the legs enables the lice to cling to hairs; specimens are often collected still attached to a hair (Figure 4.4).
- Males have dark coloured (sclerotized) genital structures in the middle of the posterior ventral surface (Figure 4.1b).
- Females have sclerotized gonopods on the ventral surface of the last visible segment of the abdomen (Figure 4.1c).
- Anoplura parasitize mammals, and most mammals have species of lice which specialize in feeding on them.

Human body and head lice – *Pediculus*

Structure

- The body is elongate and the abdomen is curved out at each segment.
- There are paratergal plates on the segments of the abdomen.

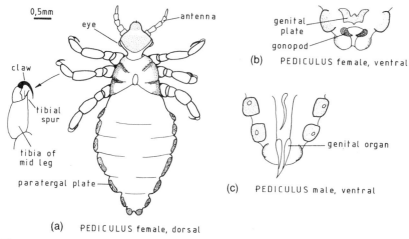

(a) PEDICULUS female, dorsal
(b) PEDICULUS female, ventral
(c) PEDICULUS male, ventral

Figure 4.1

- The head is as wide as it is long, it has antennae with five segments and a pair of well developed eyes.
- The tibial spurs have small bristles.

Other features and disease

The head louse *Pediculus capitis* infests the head of humans and the eggs laid on the hair are known as nits. The body louse *Pediculus humanus* infests the clothing of humans, particularly those unable to change their clothing. These lice feed on the body but usually lay eggs in the clothing. Heavy infestations are known as pediculosis and cause stress; rashes may develop and chronic infestations of body lice may lead to hardening and extra pigmentation of the skin. *Pediculus humanus* transmits *Rickettsia prowazekii* which causes louse borne or epidemic typhus; the rickettsia *Rochalimaea quintana* which causes trench fever; and the bacterium *Borrelia recurrentis* which causes epidemic relapsing fever. It is important to note that recent evidence confirms that the head louse, *P. capitis*, is a true separate species and that it does not transmit any pathogen. It is this head louse that is commonly found infesting schoolchildren, not the body louse.

Human pubic louse – *Pthirus*

Structure

- The body is very wide across the thorax and the abdomen is narrower and short.
- The abdomen has spiracles on the dorsal surface and three pairs of tubercles on the ventral surface.

134 LICE – PHTHIRAPTERA

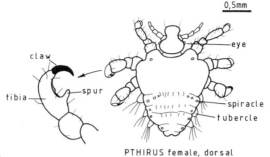

Figure 4.2

- The head is blunt at the tip and there are small eyes.
- The forelegs are thin. The mid- and hindlegs are stout and have large claws which close onto a large spur on the tibia.

Other features and disease

Pthirus pubis is also known as the crab louse because of its appearance. It infests pubic hair but also may infest armpits or beard. Irritation can be intense.

Haematopinus

Structure

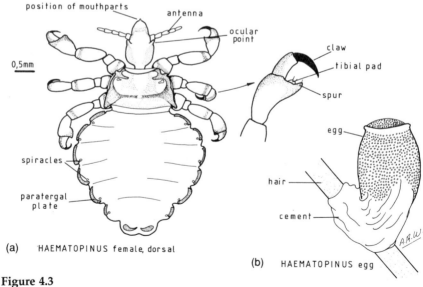

Figure 4.3

- These are large lice with distinct patterns of hardened (sclerotized) structures.
- The head is elongate, there are no eyes but behind the antennae are ocular points.
- The legs are similar in size and have large claws which close onto a large spur or hook on the tibia.
- Next to the tibial spur is a tibial pad.
- The abdomen is curved out at each segment and has paratergal plates on segments 2 or 3 through to 8.
- On the ventral surface of the thorax is a dark coloured sternal plate.

Other features and disease

Haematopinus suis, the hog louse, infests pigs, causing stress and loss of production. It is the only species of lice found on pigs. *Haematopinus eurysternus*, the short nosed cattle louse, is found all over the body of cattle and can cause stress and loss of production. *Haematopinus quadripertusus*, the cattle tail louse, and *H. tuberculatus*, the buffalo louse, are both found as minor infestations on cattle. Equids are infested by *H. asini*.

Linognathus

Structure

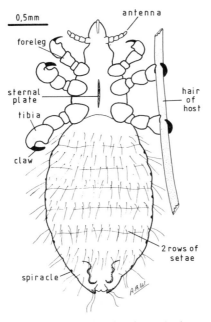

LINOGNATHUS female, ventral

Figure 4.4

136 LICE – PHTHIRAPTERA

- These are medium sized lice.
- There are no eyes or ocular points; the antennae have five segments.
- The forelegs are small. The mid and hindlegs are larger with a large claw and tibial spur but this spur is less distinct than in other Anoplura.
- The coxae of the forelegs are far apart.
- The abdomen body wall is thin and semi-transparent. It is without paratergal plates, and there are two rows of setae on each abdomen segment.
- The spiracles are indistinct.
- The sternal plate on the ventral surface of the thorax is narrow or absent.

Other features and disease

The long nosed cattle louse, *Linognathus vituli*, and the blue louse of sheep, *L. ovillus* infest the main body and head of their hosts. *Linognathus pedalis* infests the feet of sheep; *L. setosus* infests dogs, and *L. stenopsis* infests goats. Heavy infestations can cause stress and loss of production and hides for leather making are downgraded due to deformities and discolouration at the feeding sites of the lice.

Solenopotes

Structure

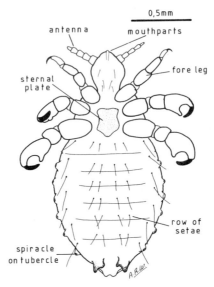

SOLENOPOTES female, ventral

Figure 4.5

- These are small lice without eyes or ocular points and having antennae comprised of five segments.
- The forelegs are small; the hindlegs are large with large claws and tibial spurs.
- The abdomen has no paratergal plates but the spiracles are on tubercles which protrude from the abdomen. There is one row of setae on each abdomen segment.
- There is a sternal plate on the thorax.

Other features and disease
The little blue louse of cattle, *Solenopotes capillatus*, infests the head and neck.

Polyplax
Structure

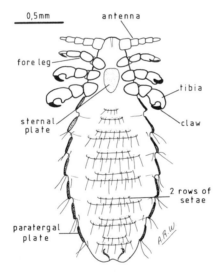

POLYPLAX female, ventral

Figure 4.6

- These lice have prominent antennae with five segments, no eyes and no ocular points.
- The forelegs are small, the hindlegs are large with large claws and tibial spurs.
- There are paratergal plates and the females have two rows of setae on the abdominal segments.
- There is a sternal plate on the ventral surface of the thorax.

Other features and disease
These are lice of rodents and may cause problems in laboratory colonies.

4.3 ISCHNOCERAN CHEWING LICE

These chewing lice have mouthparts in the form of distinct mandibles on the ventral surface of the head. The lice grasp onto hairs with the mandibles which close horizontally (Figure 4.7). They feed on skin scales and other material at the surface of the skin. They have no sucking mouthparts and there are no palps at the mouthparts. The antennae have three to five segments and the thorax is fused into a single unit. The legs end in either one or two claws depending on the species. Different species infest both mammals and birds.

Damalinia (Bovicola, Trichodectes)

The classification of *Damalinia* is complicated by use of the genus names *Bovicola* and *Trichodectes* for some species. For simplicity, *Damalinia* is used here as the genus name to include species otherwise called: *Bovicola bovis* or *Trichodectes bovis*, *B. ovis* or *T. ovis*, *T. caprae*, *B. equi* or *T. equi*.

Structure

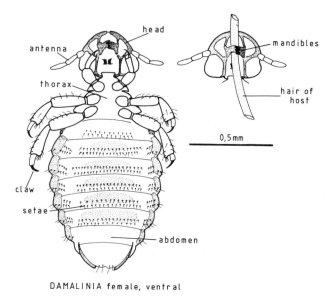

Figure 4.7

- These are small pale coloured lice.
- The head is rounded and antenna have three segments.
- The legs have a single claw.
- The abdomen has spiracles on abdominal segments 2 to 6 and setae of medium length.
- Antennae have three segments.

Other features and disease
Damalinia species infest cattle, sheep, goats and horses. Usually they cause little harm and only occur in small numbers but if large infestations develop damage may occur through heavy grooming by the host.

Trichodectes
Structure

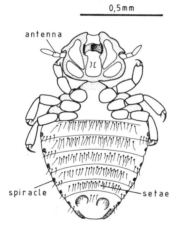

Figure 4.8   TRICHODECTES female, ventral

- The head is rounded and antennae have three segments.
- The legs end in single claws.
- The abdomen has spiracles on segments 2 to 6, and many large thick setae.

Other features and disease
Trichodectes canis infests dogs. This louse species transmits the tapeworm *Dipylidium caninum*. Other species of this genus are found on other canids.

140 LICE – PHTHIRAPTERA

Felicola

Structure

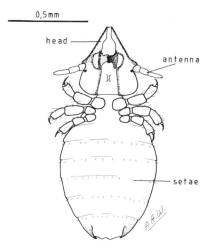

FELICOLA female, ventral

Figure 4.9

- The head is elongated in the anterior direction into a triangular shape
- Antennae have three segments.
- The legs are small and end in single claws.
- The abdomen has only three pairs of spiracles and is smooth, with few setae.

Other features and disease
Felicola subrostrata infests cats. It may become a nuisance in old or sick hosts.

Goniocotes

Structure

- This is a small compact louse, with a head lacking prominent angles.
- There are two large bristles on each side of the back of the head.
- Antennae have five segments.

Other features and disease
Goniocotes gallinae, the fluff louse, occurs on the down feathers in most parts of the body of chickens.

ISCHNOCERAN CHEWING LICE **141**

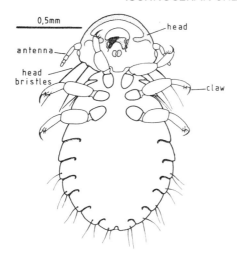

Figure 4.10 GONIOCOTES female, ventral

Goniodes

Structure

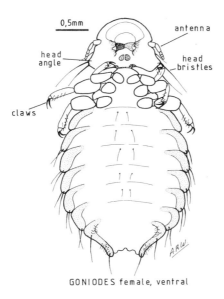

GONIODES female, ventral

Figure 4.11

- This is a large louse.
- The head has prominent angles and a distinct hollow margin posterior to the antennae.
- There are two large bristles on each side of the back of the head.
- Antennae have five segments.

142 LICE – PHTHIRAPTERA

Other features and disease
Goniodes dissimilis and *G. gigas* are found on the body and feathers of poultry.

Lipeurus

Structure

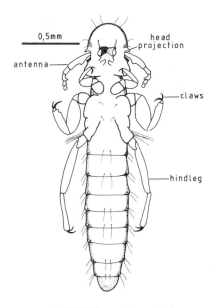

LIPEURUS male, ventral

Figure 4.12

- This is a long narrow louse.
- The head has small angular projections in front of the antennae.
- The antennae have five segments and in males these are of complex shape.
- The hindlegs are longer than the others.

Other features and disease
Lipeurus caponis is a common louse of poultry, found on the wing feathers.

Cuclotogaster

Structure

- This is a rounded louse of medium size.
- The head is rounded and without any sharp angles.

AMBLYCERAN CHEWING LICE

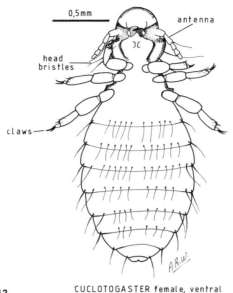

Figure 4.13 CUCLOTOGASTER female, ventral

- There are three long bristles projecting from each side of the back of the head.

Other features and disease
Cuclotogaster heterographus, the head louse, is found on the skin and feathers of the head and neck of poultry.

4.4 AMBLYCERAN CHEWING LICE

These chewing lice are superficially similar to the Ischnocera because they have mouthparts consisting of distinct mandibles on the ventral surface of the head. However the mandibles close vertically, not horizontally as in the Ischnocera. There are other more distinct differences from the Ischnocera. The Amblycera have small palps of four segments next to the mandibles. The thorax is divided between the first segment (prothorax) and the fused second and third segments (mesothorax and metathorax). The antennae have four or five segments which are folded down under the ventral surface of the head. The tarsi have one or two claws at the end. The number of claws on the tarsi is variable between species.

Menacanthus

Structure

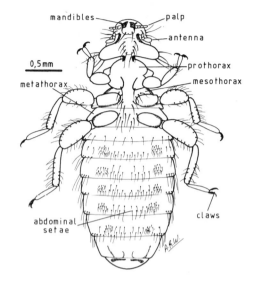

Figure 4.14

MENACANTHUS female, ventral

- A large louse with distinct palps and antennae.
- Antennae have four segments.
- The abdomen has a dense covering of medium length setae.

Other features and disease
Menacanthus stramineus, the chicken body louse, is common on poultry throughout the world.

Menopon

Structure
- A small louse, with small palps and antennae.
- Antennae have four segments.
- The abdomen has a sparse covering of medium length setae.

Other features and disease
Menopon gallinae, the shaft louse of poultry, occurs on the thigh and breast feathers.

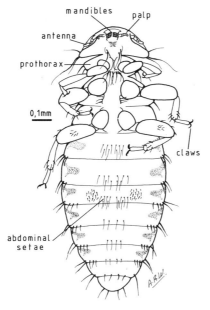

Figure 4.15

Heterodoxus

Structure

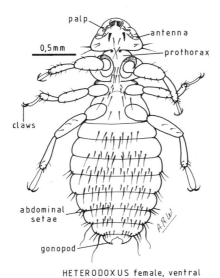

Figure 4.16

- A large louse, with a dense covering of thick, medium and long setae.
- Antennae have four segments.
- The female has distinct gonopods.

Other features and disease
Heterodoxus spinigera infests dogs and has a worldwide distribution.

5 FLEAS – SIPHONAPTERA

5.1 INTRODUCTION TO THE FLEAS

The fleas are a very distinctive group of insects with approximately 1500 species in 58 genera grouped in 15 families. They are specialized as micropredators or ectoparasites. The adult is flattened laterally and has backward pointing setae, bristles and combs of spines. This enables the flea to crawl between the hairs of its host and reduces the ability of the host to groom it away. The complex mouthparts are specialized for blood feeding by piercing the skin. The antennae are small and packed into a backward facing groove. There are no wings.

The life cycle has a complete metamorphosis and fleas are adapted for life within the housing or nests of their hosts. Eggs are laid in the nest and the larvae (Figure 5.1c) hatch out and feed on dead organic matter in the nest. Food of the larvae may include faeces of the adults which contain much partially digested blood. The larva changes into a pupa and then either a female or male adult emerges. The sexes are similar (Figure 5.1 a and b), except for the genital structures. At the end of the abdomen both sexes have a sensilium (pygidium) with numerous sensory setae. In addition the males have clasping organs of which the basimere is visible.

Both sexes feed on blood. Most fleas feed on mammals, a few species are adapted to feed on birds. The adults usually live in the housing or nest of the host and jump onto the host to feed as micropredators. In this they are similar to the blood sucking Hemiptera. Some species such as the sticktight flea *Echinophaga gallinae* remain on the host as ectoparasites and the female chigger flea *Tunga penetrans* becomes an endoparasite when feeding.

There are three main families of fleas of importance to health: the Pulicidae containing *Pulex, Xenopsylla, Ctenocephalides* and *Echidnophaga*; the Ceratophyllidae containing *Ceratophyllus* and *Nosopsyllus*; and the Tungidae containing *Tunga*. In addition, *Spilopsyllus* in the Pulicidae is of importance to rabbits in Europe where *S. cuniculi* transmits the myxoma

148 FLEAS – SIPHONAPTERA

virus and this flea is sometimes found on dogs or cats that have been hunting rabbits. The family Leptopsyllidae is represented by the mouse fleas *Leptopsylla* which may be involved in plague. There are numerous other families containing species that may be involved in plague and other diseases. It is not practical to give details here.

The most important feature for identification to genus are the combs of spines (ctenidia) on the head (genal combs), and on the first segment of the thorax (pronotal combs). These combs are conspicuous, but some other features may be only properly visible in specimens that have been made semi-transparent for mounting on a microscope slide.

Unlike lice, fleas are not highly adapted for feeding on one species of host. Usually they have a preferred species of host but will often feed on other host species. This make identification more difficult.

> *Identification to species requires specialist knowledge. Caution should be taken in identifying to genus when details are given for only a few of the many genera, as in this book.*

Cat and dog fleas – *Ctenocephalides*

Structure

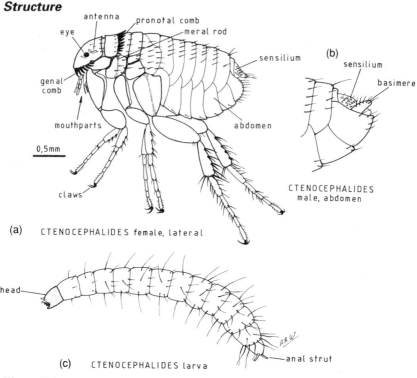

Figure 5.1

INTRODUCTION TO THE FLEAS

- The genal comb is present, with five or more spines. These spines are sharp and the posterior ones point backwards in a horizontal position.
- The pronotal comb is present.
- Eyes are present.
- There is a vertical thickening (meral rod or pleural rod) in the mesopleuron.
- The head is rounded on its upper and anterior surface. *Ctenocephalides canis*, the dog flea, has a sharp curve here; *C. felis*, the cat flea, has a shallow curve.

Other features and disease

The cat flea *C. felis* and the dog flea *C. canis* are both found on domestic cats and dogs. *Ctenocephalides felis* has become very widespread because of this association and has become a common parasite of humans. *Ctenocephalides felis* can build up to massive infestations affecting goats, leading to anaemia and severe loss of condition. The tapeworm of dogs, *Dipylidium caninum*, and the filarial worm *Dipetalonema reconditum* are transmitted by *Ctenocephalides* species.

Rat flea – *Xenopsylla*

Structure

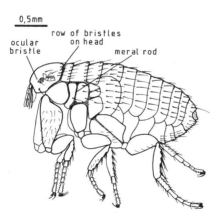

Figure 5.2   XENOPSYLLA female

- There is no genal comb or pronotal comb.
- The outer margin of the head is smoothly rounded and there is a pair of eyes.
- The thorax has large segments.
- This genus is very similar to *Pulex* but can be distinguished by the following features: a row of bristles along the rear margin of the head; the

ocular bristle is in front of the eye; in specimens that are semi-transparent on a microscope slide a vertical thickening (meral or pleural rod) in the mesopleuron of the thorax can be seen.

Other features and disease

The Oriental rat flea *Xenopsylla cheopis*, *X. brasiliensis*, *X. astia* and other species feed on domestic rats (species of *Rattus*), and as a result they have a very wide distribution throughout the world. These fleas, particularly *X. cheopis*, transmit the bacterium *Yersinia pestis* which causes plague in humans. Other genera of fleas may carry *Y. pestis* but *X. cheopis* and other *Xenopsylla* species are the most important vectors because they will readily feed on humans if their rat hosts die. Other genera of fleas involved in transmission of plague are the human flea *Pulex*, and the rodent fleas *Leptopsylla* and *Nosopsylla*.

Xenopsylla cheopis also transmits *Rickettsia typhi* causing murine or endemic typhus in humans. Other species of flea, also lice and mites, may be involved in transmission of this pathogen.

Human flea – *Pulex*

Structure

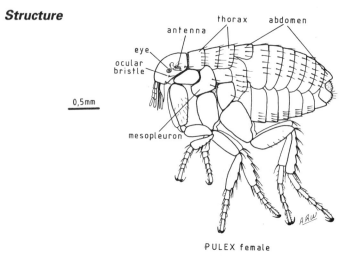

PULEX female

Figure 5.3

- There is no genal comb or pronotal comb.
- The outer margin of the head is smoothly rounded and there is a pair of eyes.
- This genus is very similar to *Xenopsylla* but can be distinguished by the following features: the ocular bristle is below the eye and there is no row of bristles along the rear margin of the head; in specimens that are

semi-transparent on a microscope slide no vertical thickening (meral or pleural rod) in the mesopleuron of the thorax can be seen.

Other features and disease
The human flea *Pulex irritans*, feeds on humans and other mammals including dogs and pigs and its distribution follows these species closely throughout the world. The human flea may become involved in transmitting *Yersinia pestis*, causing plague. There may be transmission between humans or even between rats and humans in the unusual case of this species of flea feeding on rodents.

Sticktight flea – *Echidnophaga*
Structure

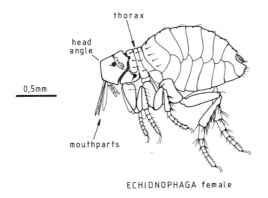

ECHIDNOPHAGA female

Figure 5.4

- There is no genal comb and no pronotal comb.
- The head is angular at the front margin and the thoracic segments are narrow at the top.
- The biting mouthparts are large and project from the head conspicuously.

Other features and disease
The sticktight flea *Echidnophaga gallinacae* is a common ectoparasite of poultry and also feeds on a wide variety of mammals including humans. The adult fleas remain on the host, and infestations on poultry occur mostly around the head and other areas of bare skin. Severe irritation and even blindness of poultry may result.

Rat flea and chicken flea – *Nosopsyllus* and *Ceratophyllus*

Structure

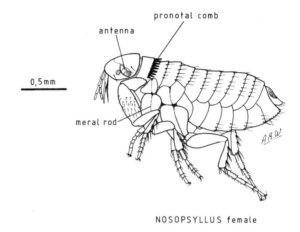

NOSOPSYLLUS female

Figure 5.5

- These genera are very similar in structure.
- The body is long.
- The areas of the head in front of and behind the antennae have less than three rows of hairs.
- There is a vertical thickening (meral or pleural rod) in the mesopleuron.
- There are no genal combs but pronotal combs are present.
- In *Nosopsyllus* the pronotal comb has 18 to 20 spines, in *Ceratophyllus* it has more than 24 spines.

Other features and disease

Nosopsyllus fasciatus feeds on rodents and may become involved in the transmission of *Yersinia pestis* causing plague in humans. *Ceratophyllus gallinae* is commonly known as the European chicken flea but will feed on many hosts including humans.

Chigger flea – *Tunga*

Structure

- *Tunga penetrans* is known as the chigger, jigger, chigoe or sand flea. It should not be confused with chigger mites (trombiculids).
- When the female is unfed the shape is of a normal flea.

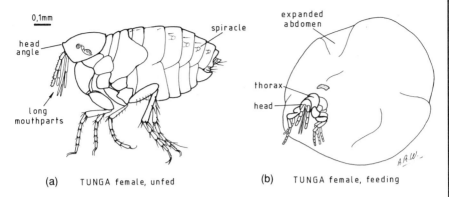

(a) TUNGA female, unfed   (b) TUNGA female, feeding

Figure 5.6

- The head is angular, it has no comb of spines but eyes are present and the mouthparts are large.
- The thoracic segments are narrow at the top.
- The female feeds by burrowing into the skin of its host. The abdomen becomes enormously enlarged between the second and third segments so that the flea forms a round sack with the head and thorax at one end and the posterior abdomen at the other end.

Other features and disease

Tunga penetrans infests humans causing great irritation. They usually invade the toes but any part of the body may be invaded. The position in the skin is in the upper dermis, with a small open wound through the epidermis. *Tunga* may also feed on pigs. The distribution is in warm humid climates, mostly in the Afrotropical Region but also in the Neotropical Region.

6 BLOOD SUCKING BUGS – HEMIPTERA

6.1 INTRODUCTION

The insect order Hemiptera consists of a very wide variety of species, many of which are common and may be pests of crops. The name bug, or true bug, is used in a technical sense for the hemipterans. The name Hemiptera refers to the wings. There are two pairs of these, with the front pair partly hardened, but some species have no wings. All species have mouthparts in the form of a proboscis (or rostrum). These consist of the fine piercing mandibles and maxillae which are sheathed in the labium. The proboscis appears segmented and is kept folded under the head when not in use. Most Hemiptera feed on plant juices but two groups have adapted to feeding on blood. These are the triatomine bugs of the family Reduviidae, and the bed bugs of the family Cimicidae.

The Hemiptera have an incomplete metamorphosis. In the blood sucking species the eggs are laid on materials in the nest or housing of the host. A nymph hatches from the egg; this resembles the adult except that it has no wings. It feeds as the adult does. There are five nymphal stages, with moulting between them. The adults are the final sexually mature forms. Adults of triatomine bugs have wings but adult bed bugs remain without wings. The blood sucking Hemiptera live in the housing or nests of their hosts. They do not remain on their hosts but feed as micropredators. In this they are similar to the fleas.

6.1.1 Triatomine bugs

The triatomine bugs are in the sub-family Triatominae. They are known as cone-nose bugs and by many other local names. There are 118 species in 15 genera, of which the genera of most medical importance are:

156 BLOOD SUCKING BUGS – HEMIPTERA

Triatoma, Panstrongylus and *Rhodnius*. These are similar and general features are given below, followed by features to distinguish the genera.

General Structure

- The adult body is long and the abdomen is broad and extended in a rim (connexivia) at the sides. The abdomen becomes rounded after feeding.
- Two pairs of wings are present. When not in use they are folded closely together over the abdomen.
- The forewings (hemelytra) are slightly hardened at the base but the remaining areas are transparent.
- The thorax has a broad pronotum and from the back of this projects a long thin scutellum.
- The head is prominent, with medium sized eyes, a pair of ocelli, and long antennae with four segments.
- The long proboscis (rostrum) lies underneath and parallel to the head, it has the appearance of being in three segments. The proboscis extends forward when in use.

Triatoma

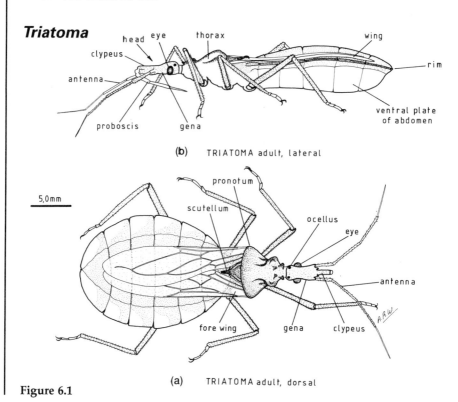

Figure 6.1

- The head is long. The clypeus at the front of the head is narrow at the top and the genae (lateral sides of the head) diverge toward the front of the head.
- The scutellum is downturned.
- The ventral plates of the abdomen do not extend to the rim or lateral margin of the abdomen.
- The base of the antenna is at the middle of the area in front of the eyes.
- The legs are all similar and have tarsi of three segments and with medium size double claws.

Rhodnius

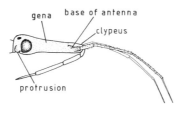

RHODNIUS head

Figure 6.2

- The general features are similar to *Triatoma*.
- The distinctions are on the head. This is very long. The clypeus at the front of the head is broad at the top, the genae are parallel. There is a protrusion (callosity) behind each eye and the base of the antenna is at the front of the head away from the eyes.

Panstrongylus

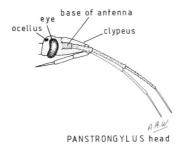

PANSTRONGYLUS head

Figure 6.3

- The general features are similar to *Triatoma*.
- The distinctions are on the head. This is short and broad and the base of the antenna is just in front of the eyes.

Sexes
The sexes are similar and both feed on blood.

Distribution
The genera which are medically important are found in the tropical humid, warm humid and dry climates of the Neotropical Region; but within this region they may be found in areas not typically tropical such as above 3000 m altitude and south of 45° latitude.

Hosts
The domestic species which are medically important feed on humans and associated domestic animals.

Life cycle
The life cycle has an incomplete metamorphosis, with five nymphal stages before the winged adults develop. All these stages live in the housing of their hosts and feed on blood.

Behaviour
Feeding is usually in the night and the bites are usually painless. Feeding takes minutes and the bugs then return to their resting places in cracks in the walls and roof of the building. Some triatomine bugs cover themselves with debris to hide. Infestations of triatomine bugs are closely associated with poor quality housing in which there are many possible resting places for the bugs.

DISEASE
HUMANS Triatomine bugs transmit the protozoan *Trypanosoma cruzi* which causes Chaga's disease. There are many species which transmit, the most important ones are: *Triatoma infestans, T. dimidiata, Panstrongylus megistus* and *Rhodnius prolixus*. There are many other species of some importance in transmission of Chaga's disease.

The bites are not usually painful but massive infestations in a house may lead to the inhabitats suffering significant blood loss. The closely related assassin bugs, which are hemipterans, may cause painful bites in self defence.

DOMESTIC ANIMALS Many other mammals are important reservoirs of infection with *Trypanosoma cruzi*, but triatomine bugs are not usually of importance to domestic animals.

6.1.2 Bed bugs – *Cimex*

Structure

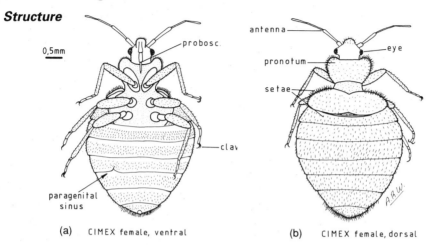

(a) CIMEX female, ventral
(b) CIMEX female, dorsal

Figure 6.4

- The adults as well as the nymphs are without wings.
- The body is flat with a rounded outline from above. The pronotum of the thorax extends laterally as a thin rim.
- The head has a four segmented antennae and eyes.
- The proboscis (rostrum) appears to be three segmented and it lies flat under the head when not in use.
- The legs are all similar and have tarsi of three segments and medium sized double claws.
- The body is covered in setae that are stout, short and blunt.
- The tropical bed bug, *Cimex hemipterus*, is shown in Figure 6.4. The common bed bug, *Cimex lectularius*, has a pronotum which extends out further laterally and posteriorly.

Sexes

These are similar and both feed on blood. The male has a genital organ (the paramere) extended as a curved rod from the last abdominal segment. The female has an indentation (paragenital sinus) on the ventral margin of the fifth abdominal segment.

Distribution

Bed bugs have become closely associated with humans and are found very widely throughout the world. They are commonest in housing of poor quality or poor hygiene. They also occur in poultry houses.

Hosts
Humans and chickens are the only hosts.

Life cycle
The life cycle has an incomplete metamorphosis with the nymphs being similar to the adults in structure and feeding methods.

Behaviour
Bed bugs are highly domestic, they live in cracks in beds and other furniture including mattresses and cushions, and in walls. They feed at night on their sleeping hosts.

Disease
HUMANS *Cimex* can be a substantial nuisance, the bites are not usually painful during feeding but become painful later. Massive infestations can lead to significant blood loss. *Cimex* are not important as transmitters of disease pathogens to humans. Another species of bed bug which causes significant nuisance is *Leptocimex boueti* in Africa.

DOMESTIC ANIMALS *Cimex* can be a substantial nuisance in poultry houses.

Part Three
OTHER ARTHROPODS AND METHODS

7 OTHER HARMFUL ARTHROPODS

In addition to the parasitic arthropods there is a wide variety of arthropods which are directly harmful to humans and sometimes to domestic animals. The worst are the venomous species; records of scorpion and spider bite cases indicate there are more serious cases of poisoning due to them than snake bites. However, because the problem centres on emergency treatment for poisoning, these arthropods need a separate book together with other venomous animals. Photographs of many venomous arthropods are given in *A Colour Atlas of Clinical Medicine* by W. Peters.

The main organisms are listed below. There are also various other pests, parasites and transmitters of disease agents listed. Some simple drawings are given.

7.1 SCORPIONS

Figure 7.1   SCORPION

These are in the class Arachnida, order Scorpionida; the sting in the tail is highly toxic in some species. The worst risks are with the following genera: *Centruroides* in Mexico and southern USA; *Tityus* in Brazil and Trinidad; *Androctonus, Leiurus, Hottentotta, Buthacus* and *Buthus* in North Africa and the Middle East; *Parabuthus* in Africa south of the Sahara; *Mesobuthus* in south and central Asia.

164 OTHER HARMFUL ARTHROPODS

7.2 WHIPSCORPIONS

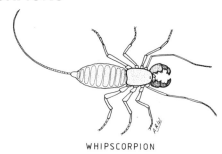

Figure 7.2

These are in the class Arachnida, order Pedipalpa; they have irritant defensive secretions but are less dangerous than they appear.

7.3 SPIDERS

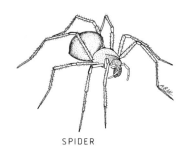

Figure 7.3

These are in the class Arachnida, order Araneida; the mouthparts can produce a poisonous bite. Many of the very large spiders known as tarantulas are not harmful to humans, but of this type the genera *Lycosa*, *Aphonopelma*, *Sericopelma* and *Pamphobeteus* are important in rainforest areas of South America and Africa.

- **Widow spiders** – *Latrodectus* of numerous species are widespread in warm climates and have a poisonous bite.
- **Recluse** or **violin spiders** – *Loxosceles* bites cause necrotic damage.
- **Wandering spider** – *Phoneutria* in South America has a poisonous bite.
- **Funnel web spider** – *Atrax* in Australia and New Zealand has a poisonous bite.

7.4 SOLPUGIDS

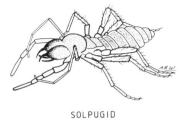

Figure 7.4 SOLPUGID

These are in the class Arachnida, order Solpugida. They appear dangerous, but are not usually harmful.

7.5 MILLIPEDES

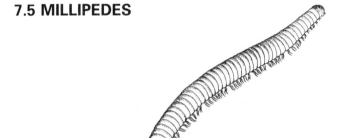

Figure 7.5 MILLIPEDE

These are in the class Diplopoda. Some tropical species have defensive secretions which cause painful skin damage with discolouration of the skin. The secretions are produced from pores along the length of the body.

7.6 CENTIPEDES

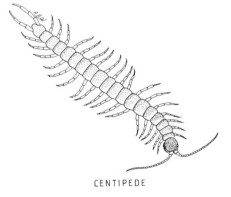

Figure 7.6 CENTIPEDE

166 OTHER HARMFUL ARTHROPODS

These are in the class Chilopoda. Some tropical species have poison claws behind the mouthparts which can deliver a painful sting.

7.7 BEES

These are insects of the order Hymenoptera. The potentially dangerous species have a poisonous sting at the end of the abdomen. The domestic honey bee, *Apis mellifera* of various sub-species, can form dangerously aggressive swarms, particularly the African form. Recent introduction of the African honey bee to South America for interbreeding with European honey bees has resulted in a dangerously aggressive form which is spreading north. Some people become hypersensitive to bee stings such that a few stings are very dangerous.

7.8 WASPS and HORNETS

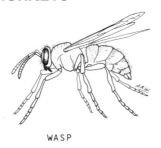

Figure 7.7

These are insects of the order Hymenoptera. The potentially harmful species have a poisonous sting at the end of the abdomen. They do not form large swarms but species of the genera *Polistes*, *Vespa*, *Vespula* and *Dolichovespula* can cause serious nuisance.

7.9 ANTS

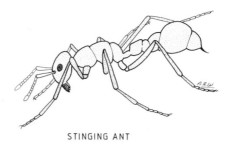

Figure 7.8

These are insects of the order Hymenoptera. Most species are harmless. Some groups have stings like bees, others have powerful and irritant bites. Fire ants, *Solenopsis* species, and harvester ants, *Pogonomyrmex* species, have severe stings. Swarms of various ants with powerful bites, known as soldier ants, are dangerous to young children.

7.10 BEETLES

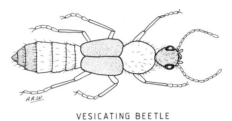

VESICATING BEETLE

Figure 7.9

These insects are in the order Coleoptera and the vast majority are harmless. The blister beetles, family Meloidae, have defensive secretions which cause skin blistering (vesication). The rove beetles, family Staphylinidae, particularly *Paedurus* genus, have blistering and irritant effects if crushed. The mealworm beetle, *Alphitobius diaperinus* is involved in the transmission of the virus of avian leucosis in poultry houses.

7.11 CATERPILLARS OF MOTHS AND BUTTERFLIES

URTICATING CATERPILLAR

Figure 7.10

These are the larvae of insects in the order Lepidoptera. Many species have defensive hairs containing poison, or defensive secretions. These will cause irritation of the skin, often severe, and also cause urticarial inflammation of the skin. They can be dangerous to the eyes. The hairs of some species can cause haemorrhage.

168 OTHER HARMFUL ARTHROPODS

7.12 ADULT MOTHS

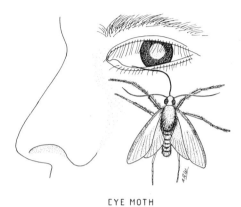

EYE MOTH

Figure 7.11

These insects are in the order Lepidoptera. They have mouthparts as a long sucking tube, the proboscis, which is usually used for feeding on plant secretions. There are some tropical species adapted to feed on tear secretions of mammalian eyes, these are known as eye moths. Other species are capable of piercing mammalian skin to feed on blood. The scales of adult moths can in some cases cause skin irritation and urticaria.

7.13 FRUIT FLIES

FRUIT FLY

Figure 7.12

These are dipteran (acalyptrate) flies in the sub-order Cyclorrhapha. Flies of the genus *Drosophila* are very common, with the larvae living in rotten fruit and vegetable matter including dung. Large numbers of adult flies may accummulate in housing of poultry, pigs and cattle and cause much nuisance to the animals and their attendants.

7.14 THRIPS

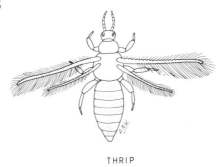

THRIP

Figure 7.13

These are insects of the order Thysanoptera. They are very small, usually winged and usually feed on plant sap using a piercing proboscis. Sometimes they swarm in huge numbers and then may cause nuisance to humans by biting.

7.15 CRUSTACEANS

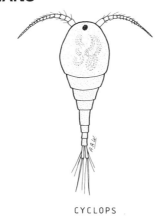

CYCLOPS

Figure 7.14

These arthropods are in the class Crustacea. Some of them act as vectors of helminth parasites. The crustaceans known as water fleas of the genus *Cyclops* and similar types, which occur in fresh water and may be ingested in the drinking water of humans, may carry the guinea worm *Dracunculus medinensis*. The broad tapeworm *Diphyllobothrium latum* has part of its life cycle in *Cyclops*. Some species of edible crabs are vectors of trematode lung flukes, *Paragonimus*, to humans.

The crustacean parasites of domesticated fish are an important constraint on fish farming but the topic is too specialized for this book.

170 OTHER HARMFUL ARTHROPODS

7.16 TONGUE WORMS

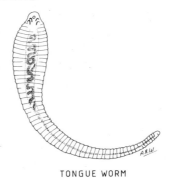

TONGUE WORM

Figure 7.15

These parasites are in the group Pentastomida, which is of uncertain status because it is not clear if they are related to the arthropods or to other animal groups. As adults they are elongated worm-like parasites. *Linguatula serrata* infects the upper respiratory tract in dogs, the viscera of rabbits and occasionally the upper respiratory tract of humans. *Armillifera moniliformis* is a parasite of snakes but may infect humans if they eat insufficiently cooked snake meat.

7.17 DELUSIONS OF INFESTATION

Some people develop delusions of infestation with arthropods (delusory parasitosis), and may produce false evidence of infestation; a similar problem is an exaggerated fear of insects and spiders (entomophobia). The resulting stress may be as great as that caused by genuine infestation. Further information on this is given in *Medical Insects and Arachnids* by Lane and Crosskey.

8 METHODS FOR IDENTIFICATION

Entomological methods are specialized for different types of arthropod and different purposes. More details are in the books given in the Bibliography with obvious titles concerning collection methods. Also useful are: Cogan, B.H. and Smith, K.G.V *Instructions for Collectors No. 4a Insects*; Smith, K.G.V. *Insects and Other Arthropods of Medical Importance*; Furman, D.P. and Catts, E.P. *Manual of Medical Entomology*; Lane, R.P. and Crosskey, R.W. *Medical Insects and Arachnids* An outline of these methods is given here. Simple drawings are given of less obvious apparatus.

8.1 COLLECTION

8.1.1 Light traps

These have an ultra-violet or visible light source which attracts flying insects. Below the light is a small suction trap, sticky trap or similar

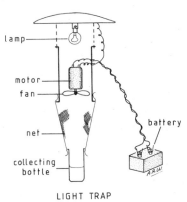

LIGHT TRAP

Figure 8.1

device to actually collect the insects (Figure 8.1). These are used for night flying mosquitoes, ceratopogonids and phlebotomines. They are about 50 cm in length and can be run from a battery and suspended from a tree or similar object in suitable habitats for insect collection.

8.1.2 Suction traps

These have a large, powerful fan to suck large volumes of air through a conical net. The net filters out flying insects. They are useful for collecting mosquitoes, ceratopogonids and other dipterans which are not attracted to light. An alternative to the suction fan is to mount the net on a vehicle and drive through areas where the insects occur. Suction traps are 1 m or more in length, they can be battery powered but often require mains electricity or a portable generator.

8.1.3 Animal baited traps

These need a host for blood-feeding which is temporarily confined in a cage or net cover. Insects attracted to the host can be suddenly enclosed by a cover. Alternatively they enter a trap around the cage, and are then collected at sites where they try to escape. These traps are useful for studies on all blood feeding flying insects, but specially those which fly in the day such as simuliids, tabanids and horn flies.

8.1.4 Bait hosts or sentinels

These are the natural blood-feeding hosts of the arthropods. They are identified by ear tag or similar means and allowed to graze with the herd, or are tethered or led through the habitat of the arthropod. The insects or ticks are then collected from the hosts using insect nets for flying insects such as tsetse or muscids, or using forceps for myiasis larvae or ticks, or an aspirator.

8.1.5 Insect nets and aspirators

These are used together with bait hosts. The net is made of mosquito netting and is small (25 cm diameter) and light enough to be quickly moved by one hand to collect tsetse, tabanids or other flies. The aspirator (Figure 8.2) is a suction tube which feeds into a collection bottle. Fine mesh cloth is fixed over the exit tube and the aspirator can be worked by an electric pump for heavy use. Aspirators are very useful for collecting a wide variety of arthropods.

COLLECTION 173

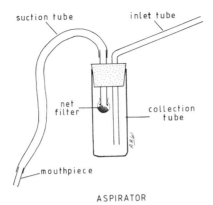

Figure 8.2

8.1.6 Sweep nets and blanket drags

These are used for collecting ticks from vegetation. The sweep net (Figure 8.3) is very strongly made with a wide (35 cm) metal frame and a long handle (120 cm) for use with both hands. The net is made from thick cloth which is also white and smooth (nylon is best). It is in the shape of a deep bag. It is brushed vigorously through vegetation to dislodge ticks. The blanket drag (Figure 8.4) is a woollen blanket or cotton towel, white in colour, about 1 m square. It is fixed to a draw bar and is pulled across low lying vegetation. Ticks attach to the cloth. In addition to the drag the operator can wear leggings of the same cloth. These are attached to the operator's trousers to collect ticks whilst walking through vegetation.

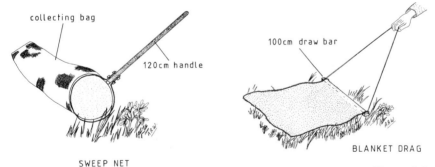

Figure 8.3 **Figure 8.4**

8.1.7 Smell baits and attractants

Substances which mimic the host odours will bring various arthropods close enough to a trap for them to be caught by baffles, suction or adhesives. Preparations of meat will attract blowflies and some other myiasis flies to lay eggs. Carbon dioxide gas released from a cylinder or dry ice attracts some biting flies to suction traps and attracts ticks on vegetation to traps with sticky collection areas. The visual traps for tsetse are often supplemented with acetone and other chemicals as odour attractants.

8.1.8 Visual traps and baffle traps

These are used for collecting actively flying insects which make much use of sight to find hosts. Tabanid flies are collected in a variety of traps which are either large arrangements of fine netting like an open sided tent (a Malaise trap) or have a dark coloured sphere freely hanging below a netting cone (the Manitoba trap). The tabanids try to escape upwards from the baffles and are led into collecting containers. Tsetse are collected in traps made of cloth and netting in the shape of two cones sewn together (biconical trap, Figure 8.5) or in box shapes. Tsetse are attracted to the shape and dark colour of the cloth and enter openings in the dark cloth. On trying to escape they fly up into the netting area and into collection containers.

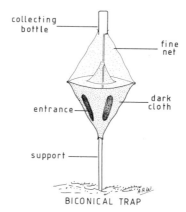

Figure 8.5

8.1.9 Electrocutor traps

These can be used to collect muscid flies or tsetse. A rectangle of parallel wires supplied with electricity at very high voltage but low current will electrocute insects that contact it. The insects are collected in a tray. This can be used for muscid flies attracted to a light or for tsetse attracted to a dark or moving object. For example they can be fitted to the back of a vehicle and tsetse attracted to the vehicle will be trapped.

8.1.10 Sticky traps

These are made using a grease such as petroleum jelly or tree banding grease or a non-drying polymer adhesive. This is spread over a sheet of plastic or wood and flying insects landing on it will be trapped. For collecting ticks from vegetation the sticky trap surrounds a central source of carbon dioxide as an attractant.

8.1.11 Emergence traps

These are placed over larval habitats such as areas of water or swampy ground or dung piles to collect adult dipterans newly emerging from their larval habitats (Figure 8.6). Alternatively samples of the habitat are placed in the trap back at the laboratory. They are used for simuliids, phlebotomines, ceratopogonids, eye flies, muscid flies and mosquitoes. There are also methods for separating larvae from larval habitats such as mixing the larval habitat with a dense liquid such as concentrated sugar solution or saturated magnesium sulphate. The larvae float to the surface where they can be collected.

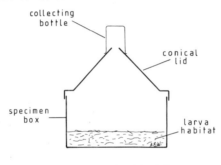

Figure 8.6

EMERGENCE TRAP

8.1.12 Host searches

These are used to collect ectoparasitic insects, ticks and mites. Fleas and lice are collected from live hosts by parting the hair or feathers and picking up the lice with forceps or an aspirator. They can also be collected by dusting the host with an insecticide and collecting the dead parasites as they fall onto a tray.

Ticks are collected from livestock by holding them in a crush or casting them to the ground. It is often convenient to search only one side or the known feeding sites. The ticks are removed with forceps by grasping the tick over the mouthparts and pulling firmly. Ticks removed from hosts often have attachment cement around their mouthparts. This should be cleaned off using fine forceps before examining the ticks.

Mites causing mange or scab need to be collected by scraping the skin surface with the edge of a microscope slide or a scalpel held at 90° to the skin. If the scraper is wetted with oil or glycerol this can be the medium for making a temporary mount on a microscope slide for examination. Sarcoptic mange mites occur within the epidermis and the scrapings need to go into the epidermis, thus drawing spots of blood. For demodectic mange mites either squeezing the affected skin or cutting into nodules will produce the characteristic white mass of mites and **sebum**.

8.2 FRESH SPECIMENS FOR PATHOGEN ISOLATION, BLOODMEAL TESTS AND AGE-GRADING

Often the arthropods collected are needed alive or preserved freshly so that they can be tested for presence of pathogenic organisms; for the source of any blood-meals they contain; or to grade them into age classes.

Dipterans such as mosquitoes, ceratopogonids, phlebotomines, tabanids or tsetse can be collected live if the collecting chamber of the trap is made of fine mesh. Preferably the chamber should be protected from the effects of direct sun and drying air and the insects removed promptly. They can be taken to the laboratory directly in the chambers which should be kept in a cool and humid box. Alternatively they can be sucked out with an aspirator and placed over water ice, dry ice or liquid nitrogen. The specimens should only be preserved by freezing if it is known that the method for isolating pathogens will work with frozen material.

Ticks should be placed in tubes with a ventilated stopper or a cotton wool plug covered in gauze. The tube should contain paper for the ticks to hold onto and a label written on paper. The tubes are kept in a cool and humid box for transport back to the laboratory. This box should be insulated, and contain ice blocks or freezer blocks and wet pads to maintain high humidity. The ticks should not come in direct contact with the ice. A large vacuum flask can be used.

8.3 PRESERVATION

8.3.1 Liquid preservative

Mites, ticks and ectoparasitic insects are usually collected into ethanol (ethyl alcohol) which is diluted to 75% ethanol by addition of water. A useful additive is glycerol, at one part in ten of the total; this prevents accidental drying out of specimens. Formalin at 2% to 10% in water can be used. It may be easier to obtain and use in the field; it does not evaporate as rapidly as ethanol but it is unpleasant and possibly dangerous to use. Field specimens in formalin can be transferred to ethanol for laboratory work. Temporary preservation can be obtained in industrial alcohol, medical methylated spirit or similar alcoholic liquids.

Live myiasis larvae shrink if placed directly in ethanol. A mixture of three parts of 90% ethanol plus one part of glacial (concentrated) acetic acid gives a preservative which will not cause shrinkage. This is useful for field work but in the laboratory it is simplest to kill the larvae in water at about 80 °C and then preserve them in the usual 75% ethanol.

The best containers are the thick-walled glass specimen tubes known as Bijou bottles in the 5 ml size or Universal bottles in the 25 ml size. They have a wide mouth, with metal screw caps and a rubber washer and are very resistant to breakage. Few other containers are suitable. Plastic containers should only be used for field collection as they leak during long storage. A temporary label should be written with pencil on paper and placed **inside** the tube. After examination a label should be written in waterproof carbon ink (Chinese or Indian ink) on thin cardboard, to be placed permanently inside the tube. The label should have details of site and date of collection, and the collector's name.

Do not give a species name unless the specimen has been positively identified to species level.

When arthropods are placed in alcohol they tend to curl up, making identification difficult. Under laboratory conditions this can be avoided by killing fresh specimens in water at about 80 °C; they are then transferred to preservative.

8.3.2 Dry preservation

The adults of larger insects such as tsetse or blowflies can be preserved dry; this has the advantage of preserving the setae, spines and colour better for identification. Mosquito adults are not easy to handle for dry preservation but this may be essential to retain the patterns of scales and setae for identification. The insects are killed in the field by freezing or in a jar containing a pad wet with chloroform, ethyl acetate or carbon tetrachloride. The liquid killing agent must be kept separate from the insects

Whilst the insects are still soft they are fixed on a pin through the thorax to one side of the median line so that one set of characters is undamaged. The legs and wings are arranged in a convenient position. The pin is also passed through a label and is fixed in a cork board in an entomological specimen box or in a jar. This is a specialized method for which entomological equipment suppliers provide the correct type of pins and boxes. Special boards are also available for setting out the specimens in the correct position.

8.3.3 Preservation on microscope slides

This is used for the mites and smaller insects such as fleas, lice, ceratopogonids and phlebotomines. These specimens can also be examined directly from preservation in liquid in a tube, but if correctly mounted on a microscope slide important features for identification may become visible. There are three basic methods: temporary mounts in water soluble **media**; permanent mounts in water soluble media; permanent mounts in water insoluble media. An outline of some simple methods is given below.

Freshly collected mites in glycerol or oil can be mounted directly on a microscope slide, a cover slip placed over and the specimen examined as a temporary mount.

For permanent mounting the specimens are first preserved (fixed) by immersion in ethanol or formalin for several hours. A convenient water soluble medium is the recipe of Berlese: dissolve 140 g of chloral hydrate in 20 ml of water, add 5 ml of concentrated (glacial) acetic acid, add 10 ml of glycerol and then dissolve 16 g of powdered gum arabic (gum acacia) into the mixture. Keep the mixture warm over a water bath to fully dissolve all components and filter it if the gum arabic makes it too cloudy. The specimens are placed in a drop of this medium on a microscope slide and a cover slip is added.

Permanent mounting in water insoluble media gives the best results. The specimens are dehydrated by soaking in several mixtures of ethanol and water, starting at 75% ethanol, up to 100% ethanol. The specimen is then clarified by soaking in xylene and it is then placed in a drop of Canada balsam medium on a microscope slide and a cover slip is added. The times spent soaking vary with size, up to one hour each. Large or dark specimens such as fleas need to be clarified before dehydration. Soak them in 10% potassium hydroxide solution for about one day, wash them in water for a few minutes, soak them in 10% acetic acid for 30 minutes, wash in water again before starting the dehydration procedure.

8.4 EXAMINATION OF SPECIMENS

Most of these arthropods need to be examined using some visual magnification. A simple hand lens of ×10 magnification is useful for field work and will permit many identification features to be checked in the laboratory.

A low power stereoscopic microscope (dissecting) is the most useful equipment for examinations. These have two separate paths for the light and prisms like a pair of binoculars so that the image is three dimensional and is upright. Several magnifications of **objective lens** betweeen ×0.5 to ×5 are useful. These are used together with an eyepiece lens, usually of ×10 magnification. A powerful and concentrated lamplight is required for these microscopes. The best light is through a fibre optic cable which avoids heating the specimens. They can be used for examination of specimens in dishes, on microscope slides or as dry pinned specimens. The latter can be viewed in any position by placing the pin at angles in a block of rubber.

Specimens preserved in liquid may be easier to examine for surface features if the surface is allowed to dry. Such specimens should not be kept out of the liquid preservative for too long. They can be rinsed in 100% ethanol and dried under a bench lamp or by using blotting paper.

For mites, a compound microscope of the sort used for examination of pathology specimens is used, but a very useful feature is phase contrast or dark ground illumination. The magnifications of the objective lens used for mites is ×10 to ×50.

Very informative and fascinating images of arthropods are made using scanning electron microscopes. These are most important for taxonomic studies but unfortunately they are very expensive, thus are not widely available for practical use in identification.

Specimens are handled using a basic tool kit as follows:

- four grades of forceps: large robust and blunt ended ones for removing ticks; medium size fine pointed ones for removing lice and fleas; very fine pointed watch-makers forceps for handling mites; fine flexible forceps made of springy steel for handling specimens in the laboratory without crushing;
- several sizes of scissors;
- fine needles mounted in holders;
- a scalpel with fine pointed disposable blades;
- plastic pipettes with rubber bulbs;
- specimen tubes, labels, with pencil and carbon ink pen for labelling;
- dishes for examining specimens;
- a killing jar for large insects to be preserved dry;
- a small supply of liquid preservative in a plastic bottle for field use.

Appendix A:

SUMMARY OF CLINICAL SIGNS OF DIRECT HARM CAUSED BY ARTHROPODS

SIGN OR DISEASE	ARTHROPOD
abscess	*Amblyomma, Hyalomma*
alopecia	*Psoroptes, Demodex*
anaemia	*Ctenocephalides*
anaphylaxis	*Simulium*
allergy (asthma, dermatitis, rhinitis	*Glycyphagus, Acarus, Dermatophagoides, Pyemotes, Simulium*
blisters	allergic reaction to blood sucking insects, ticks and mites; effect of irritants from beetles or millipedes
boils	*Dermatobia, Cordylobia, Wohlfahrtia, Hypoderma, Tunga*
depluming	*Knemidokoptes, Megninia*
dermal nodules or plaques	*Demodex*
fever	*Rhipicephalus, Simulium*
hypersensitivity (dermal)	repeated bites of most arthropods
immune suppression	*Amblyomma*
lung and airsac infestation	*Cytodites, Pneumonyssoides, Pneumonyssus, Sternostoma*
mange	*Sarcoptes, Notoedres, Demodex.*
myiasis	
furuncular	*Dermatobia, Hypoderma, Wohlfahrtia, Cordylobia*
gastric or rectal	*Gasterophilus*
nasal	*Oestrus, Cephalopina, Rhinoestrus, Gedoelstia*
ophthalmic	*Oestrus, Hypoderma, Gasterophilus*
subcutaneous	*Cochliomyia, Chrysomya*

APPENDIX A

SIGN OR DISEASE	ARTHROPOD
superficial	*Lucilia, Calliphora, Phormia, Sarcophaga*
systemic	*Hypoderma*
urinary	*Musca, Fannia*
warbles	*Hypoderma*
nausea	*Simulium* (see toxicosis)
nuisance (non-biting)	*Musca, Hydrotaea, Morellia, Hippelates*
otitis and ear canker	*Psoroptes cuniculi, Otodectes, Otobius, Raillietia, Rhipicephalus*
panic	adults of *Hypoderma*, tabanids, and *Gasterophilus*
paralysis	*Dermacentor, Ixodes, Rhipicephalus*
poisoning (acute) (see toxicosis)	spiders, scorpions, centipedes, ants, bees, wasps (see sections for typical genera)
pruritus (chronic)	*Sarcoptes, Psoroptes, Psorergates, Pediculus, Pthirus*
rash (distant from infestation)	*Sarcoptes*
scab	*Psoroptes, Chorioptes*
scaly leg	*Knemidokoptes*
stress (biting)	*Simulium, Culicoides*, mosquitoes, tabanids, *Haematobia, Amblyomma, Boophilus, Rhipicephalus, Ornithodoros, Linognathus, Haematopinus, Ctenocephalides*
strike (blowfly)	*Lucilia, Calliphora, Phormia*
subcutaneous nodules	*Laminosioptes*
sweating sickness	*Hyalomma*
toxicosis (see poisoning)	*Hyalomma, Rhipicephalus, Ornithodoros, Simulium*

Appendix B:

SUMMARY OF DISEASES AND CAUSATIVE PATHOGENS COMMONLY TRANSMITTED BY ARTHROPODS

(For other diseases, alternative names of diseases and other pathogens consult the index; the designation human or domestic animal refers to clinically important disease, it does not include reservoir hosts.)

DISEASE	PATHOGEN	DISEASE HOST		ARTHROPOD VECTOR
		Human	Domestic animal	
aegyptianellosis	*Aegyptianella pullorum*	-	+	*Argas*
anaplasmosis	*Anaplasma* spp.	-	+	*Boophilus* *Dermacentor* tabanids
anthrax	*Bacillus anthracis*	-	+	tabanids
African horse sickness	African horse sickness virus	-	+	*Culicoides*
African swine fever	African swine fever virus	-	+	*Ornithodoros*
babesiosis	*Babesia* spp.	+	+	*Boophilus* *Dermacentor* *Hyalomma* *Ixodes*
bluetongue	Bluetongue virus	-	+	*Culicoides*
borreliosis; relapsing fevers (see also Lyme disease)	*Borrelia* spp.	+	-	*Argas* *Ixodes* *Ornithodoros* *Pediculus*

184 APPENDIX B

DISEASE	PATHOGEN	DISEASE HOST		ARTHROPOD VECTOR
		Human	Domestic animal	
guinea worm	*Dracunculus medinensis*	+	-	*Cyclops*
haemorrhagic fevers	Dengue virus	+	-	*Aedes*
	Crimea-Congo virus	+	-	*Hyalomma*
heartworm	*Dirofilaria immitis*	-	+	*Aedes*
Carrion's disease (bartonellosis)	*Bartonella bacilliformis*	+	-	*Phlebotomus*
Chaga's disease	*Trypanosoma cruzi*	+	-	triatomine bugs
Colorado tick fever	Colorado tick fever virus	+	-	*Dermacentor*
Cowdriosis (see heartwater)				
East Coast fever (theileriosis)	*Theileria parva*	-	+	*Rhipicephalus*
ehrlichiosis	*Ehrlichia* spp.	-	+	*Hyalomma* *Rhipicephalus*
encephalitis	various encephalitis viruses	+	+	*Aedes* *Culex* *Culiseta* *Psorophora* *Ixodes* *Dermacentor* *Haemaphysalis*
equine infectious anaemia	E.I.A.virus	-	+	tabanid flies
filariasis	*Dipetalonema* spp.	+	+	*Culicoides*
	Mansonella ozzardi	+	-	*Culicoides*
	Wuchereria and *Brugia* spp.	+	-	*Culex* *Anopheles* *Aedes* *Mansonia*
	Loa loa	+	-	*Chrysops*
	Onchocerca spp.	+	+	*Culicoides* *Simulium*

APPENDIX B **185**

DISEASE	PATHOGEN	DISEASE HOST		ARTHROPOD VECTOR
		Human	Domestic animal	
	Stephanofilaria stilesi	-	+	*Haematobia*
	Parafilaria	-	+	*Musca*
heartwater (cowdriosis)	*Cowdria ruminantium*	-	+	*Amblyomma*
heartworm	*Dirofilaria immitis*	-	+	*Aedes*
kerato-conjunctivitis	*Moraxella bovis*	-	+	*Musca*
Kyasanur forest disease	Kyasanur forest disease virus	+	+	*Haemaphysalis*
leishmaniasis (kala azar, oriential sore)	*Leishmania* spp.	+	-	*Phlebotomus Lutzomyia*
lung worm (human)	*Paragonimus*	+	-	crabs
Lyme disease	*Borrelia burgdorferi*	+	-	*Ixodes*
malaria	*Plasmodium* spp.	+	-	*Anopheles*
Nairobi sheep disease	Nairobi sheep disease virus	-	+	*Rhipicephalus*
Nagana (animal trypanosomiasis)	*Trypanosoma*	-	+	*Glossina* spp.
plague	*Yersinia pestis*	+	-	*Xenopsylla*
Q-fever	*Coxiella burneti*	+	+	*Ixodes Dermacentor*
relapsing fevers (see borreliosis)				
rickettsiosis; typhus; spotted fever, etc.	*Rickettsia* species etc.	+	+	*Amblyomma Dermacentor Haemaphysalis Hyalomma Ornithonyssus* trombiculids

APPENDIX B

DISEASE	PATHOGEN	DISEASE HOST		ARTHROPOD VECTOR
		Human	Domestic animal	
				Pediculus
				Xenopsylla
Rift Valley fever	Rift Valley fever virus	+	+	*Aedes*
				Culex
river blindness (onchocerciasis)	*Onchocerca* spp.	+	–	*Simulium*
sandfly fever	Sandfly fever virus	+	–	*Phlebotomus*
sleeping sickness (human trypanosomiasis)	*Trypanosoma* spp.	+	–	*Glossina* spp.
spirochetosis (see borreliosis)				
tapeworm	*Dipylidium caninum*	–	+	*Cteno-cephalides*
				Trichodectes
tick borne fever	*Cytocetes phagocytophila*	–	+	*Ixodes*
trench fever	*Rochalimaea quintana*	+	–	*Pediculus*
tropical theileriosis	*Theileria annulata*	–	+	*Hyalomma*
tularaemia	*Francisella tularensis*	+	–	*Dermacentor*
				Haemaphysalis
				Ixodes
typhus (murine)	*Rickettsia mooseri*	+	–	*Xenopsylla*
typhus (scrub)	*Rickettsia tsutsugamushi*	+	–	trombiculids
typhus (tick)	*Rickettsia conori*	+	–	*Dermacentor*
	R. sibirica	+	–	*Rhipicephalus*
	R. australis	+	–	*Amblyomma*
				Hyalomma
				Ixodes
				Haemaphysalis
yellow fever	Yellow fever virus	+	–	*Aedes*

GLOSSARY

(For names of groups of organisms also check the list of Contents, Table 1 and the Index.)

Abscess A local concentration of pus resulting from inflammation around a point of infection or foreign object.

Acalyptrate Flies with no or small squamae or calypters.

Afrotropical Region The area of distribution of animals comprising Africa south of the Sahara, including Madagascar.

Allergy An immune reaction of the hypersensitivity type which occurs in response to a particular substance.

Alopecia Loss of hair, baldness.

Alula A partially separate flap of the wing of dipteran flies, situated at the base of the wing.

Amblyceran A chewing louse, of a sub-order of insects the Amblycera.

Anal groove In ticks, a shallow groove of semi-circular shape, posterior or anterior to the anus.

Anal plate A hardened (sclerotized) part of the integument of male hard ticks which protrudes from the surface when male is sexually active; may be known as anal shields or ventral plates; there are three pairs–adanal plates nearest the anus, accessory plates lateral to the adanal plates, subanal plates posterior to the adanal plates.

Anopheline A mosquito belonging to the sub-family Anophelinae, comprising the genera *Anopheles* and two rare genera *Bironella* and *Chagasia*.

Anopluran A louse belonging to the suborder Anoplura, known as the sucking lice.

Antenna A paired sensory organ of insects which protrudes from the head near the eyes and is mainly for odour perception.

Anterior Situated toward the head end of an animal.

Anus The opening of the digestive tract at the opposite end to the mouth, where excretion occurs.

Apodeme Extension along the body of the coxa of some mites, for muscle attachment.

Apterous Without wings, usually to describe wingless insects.

GLOSSARY

Apterygota A sub-class of the class Insecta; primitive insects which have never developed wings.

Argasid A tick of the family Argasidae, known as soft ticks because they do not have large areas of hardened plates on the body, they are very tough despite the name.

Arista A part of the antenna of cyclorrhaphan flies which protrudes from the third segment like a feather or bristle.

Arthropod An animal of the phylum Arthropoda, characterized by a hardened exoskeleton, segmented body and numerous paired limbs; includes insects, mites, ticks, spiders, scorpions, millipedes, centipedes, crustaceans and others.

Assassin bug An insect of the order Hemiptera, related to the triatomine bugs, may cause painful bites.

Asthma An allergic condition which affects the respiratory system, many substances cause it including products from mites, urticating hairs from moth and beetle larvae, and fragments of cast skins from the mass emergence of some aquatic insects.

Astigmatid A mite of the sub-order of Astigmata, such as scab mites.

Australasian Region The area of distribution of animals comprising Australia, New Zealand, the Pacific Islands, New Guinea and surrounding islands; the separation from the Oriental Region is at the Strait of Makassar between Sulawesi and Borneo using the system of Wallace; the system of Weber places Sulawesi in the Oriental Region.

Axilla The area of mammals and birds underneath the forelegs, arms or wings; the armpit.

Bacteria A group of microorganisms, some of which are parasitic, usually single celled and with a simple nucleus, intermediate in size between protozoa and rickettsia, for example *Yersinia pestis* transmitted by *Xenopsylla* fleas and causing plague in humans.

Basimere Part of the external reproductive apparatus of male fleas.

Basis capituli The anterior part of the body of mites and ticks from which the mouthparts project forward (*see* **Capitulum**)

Bite The informal name for any action by which an arthropod pierces the skin of its host in order to feed; more strictly a bite is made by the mandibles of the mouthparts closing towards each other to cut something.

Blowfly Informal name for flies of the family Calliphoridae, order Diptera; with larvae which usually feed on dead flesh but which may feed on live flesh causing myiasis.

Boil An inflamed swelling in the skin, either a true abscess or a parasitic warble as occurs in furuncular myiasis.

Bovid A mammal of the family Bovidae, which are ruminant animals with horns; typically sheep, goats, cattle, antelope, buffalo.

Bristle A large seta; a moveable projection like a hair from the integument of an arthropod.

Bug Name used in informal and formal sense for insects of the order Hemiptera such as triatomine bugs and bed bugs.
Calliphorid A fly of the family Calliphoridae in the order Diptera; typically the blowflies.
Callosity A hardened bump or protrusion from the surface.
Calypter Alternative name for squama; membranous flap at the base of the wings of some Diptera.
Canid A mammal of the family Canidae, typically dogs, jackals wolves, foxes.
Canker Skin infection causing inflammation and scabbing, typically caused by mites in the ear; otitis.
Capitulum The anterior body part of mites and ticks including the mouthparts; also known as the gnathosoma.
Caruncle The adhesive end part of the sucker on the leg of mites.
Caudal appendage The projection of the posterior body of some male ticks which appears when they are sexually mature after feeding; its function is not clear.
Ceratopogonid A fly of the family Ceratopogonidae, order Diptera; typically a biting midge such as *Culicoides*.
Cerci Structures at the end of the abdomen of insects, may be used in reproduction.
Chelicera A paired organ which is the main piercing apparatus of the mouthparts of mites and ticks.
Cheliceral sheath A part of the mouthparts of mites and ticks which covers the chelicerae and, together with the hypostome, forms the sucking tube used in feeding.
Chew To grind food using the mandibles of the mouthparts, as in the chewing lice.
Claspers Structures at the end of the abdomen of male insects used to couple with the femae during mating.
Class A group in the taxonomy of animals, (for example Insecta), between phylum (for example Arthropoda) and Order (for example Diptera).
Claw A structure on the feet of mites, ticks and insects in the form of a hard sharp hook, usually paired.
Clypeus A part of the head of insects, the dorsal part between the antennae and the mouthparts.
Cockle Damage to hides caused by feeding of hippoboscid flies.
Comb In some species of fleas there are distinctive rows of spines on the head or thorax which look like a comb; also called a ctenidium.
Connexivia A rim around the lateral margin of the abdomen of triatomine bugs.
Copulatory protuberance In some species of mites the pubescent female has a pair of these organs at the posterior end of the body; they connect with the copulatory suckers of the males during mating.

Copulatory suckers In some species of mites the male has a pair of these organs at the posterior end; they connect with the copulatory protuberances of females during mating.
Coxa The first segment of the legs of mites, ticks and insects, joined to the main body.
Ctenidium A row of spines on the head or thorax of fleas; shaped like and also known as a comb.
Culicine A mosquito belonging to the sub-family Culicinae; includes most mosquitoes; the examples in this book are *Culex, Aedes, Mansonia, Coquilletidia, Haemagogus, Sabethes, Psorophora* and *Culiseta*.
Cuticle The outer dead and protective layer of the integument of mites, ticks and insects.
Cyclorrhaphan A fly in the sub-order known as Cyclorrhapha, of the order Diptera, may be known as the infra-order Muscomorpha; characterized by larva forming a puparium from the last larval skin inside which pupation occurs; usually they have antenna of three segments, the outer third segment being large and having a protruding arista; typically blowflies and house flies.
Demodecosis Infestation of the skin with *Demodex* mites.
Dermatitis Any inflammatory condition of the skin, often occurs as a direct result of arthropod bites, or as an indirect effect of allergic reactions to arthropods.
Dermis The main part of the skin which occurs below the outer epidermis and contains the blood vessels, hair roots and nerves.
Desiccation Loss of body water; dehydration.
Dewlap The flap of skin on the ventral side of the neck of zebu type cattle.
Dichoptic In the Diptera, having eyes separated above by a wide gap; typical of females.
Dipteran A true fly of the order Diptera.
Diurnal Active during daylight.
Dorsal In arthropods the dorsal side of the body is the side opposite from where the legs project and make contact with the surface; in vertebrates it is the side where the spine occurs.
Ectoparasite (*see* **Parasite**).
Empodium A structure on the feet of Diptera flies, in the form of a single pad or seta betweeen the paired pulvilli and claws.
Endoparasite (*see* **Parasite**).
Endopterygota Insects which have a complete metamorphosis; a division of the sub class Pterygota (winged insects) of the class Insecta; the wings develop only within the pupa.
Engorge To feed fully, in this context usually full with blood.
Epidermis In vertebrate animals this is the outer layer of the skin containing regenerative cells and a top layer of dead protective cells but no

blood vessels; in arthropods the epidermis is the inner living layer of the integument which produces the protective cuticle.

Equid A horse or donkey or other member of the mammal family Equidae.

Exopterygota Insects which have an incomplete metamorphosis; a division of the sub-class Pterygota (winged insects) of the class Insecta; the wings develop externally on the immature stages and become fully functional in the adults.

Exoskeleton The outer body layer, also known as the integument or cuticle, of arthropods which has the muscles for locomotion attached to it internally; it functions as an external skeleton.

Facet The outer layer of one of the many separate optical units that make up the compound eyes of arthropods.

Facultative The ability to live under different conditions; a facultative parasite will usually be free-living but also able to live as a parasite.

Family A group in the taxonomy of animals, for example Muscidae; between order (for example Diptera) and genus (for example *Musca*).

Femur The segment of the legs of arthropods between the trochanter and the tibia (sometimes the trochanter is fused with the femur and this combined limb is then situated between the coxa and the tibia).

Festoon Rounded patterns or lobes in the outline of the posterior body wall of some genera of ticks.

Filariasis Infection with nematode worms, within the lymphatic system of the host.

Free-living Living in physical separation from the source of food; in contrast to being parasitic, but the distinction is not clear in the case of many blood feeding arthropods.

Furuncular myiasis Myiasis where single larvae form separate infestations below the skin, like boils.

Gadding Panic caused by adults of some flies (properly applied to warble flies, *Hypoderma*) causing myiasis, when they attempt to lay eggs on the host.

Gena An area (cheek) of the head of insects, at the side.

genal Concerning the gena (or cheek) of the head of insects, for example the genal comb of fleas.

Genital pore The opening in the body wall for the male or female reproductive apparatus.

Genus The smallest main group (assemblage of species) in the taxonomy of living organisms, for example *Musca*; a genus will contain one or more species each of which has the same genus name and its own specific name, for example *Musca domestica*; there are also sub-genera and species groups.

Gnathosoma The anterior section of the body of mites and ticks; it has the mouthparts but it is not the same as the head of insects.

Gonopod A general term for the structures on the ventral and posterior surface of the abdomen of insects that form the external genital organs; the term gonapophysis may also be used.

Groin The inner area of the back legs of vertebrate animals where they join the abdomen.

Habitat The physical environment in which a species of living organism needs to be for successful survival.

Hair In mammals hairs develop from many epidermal cells of the hair follicles and they form a protective outer layer of the skin; the term hair is also used informally for many similar structures of arthropods but the technical term for these is setae.

Hair follicle The deep pocket of epidermal tissue of mammals from which the hairs develop.

Hairy Having many hairs as do many mammals, or having many setae giving the same appearance.

Haller's organ A pit in the forelegs of ticks which contains several types of sensory organs for detecting food and mates; the forelegs are used in the same way as antennae of insects.

Haltere A paired organ in the Diptera, shaped like a club, used as a sensory aid for balance when flying; found behind the wings and formed by modification of the hindwings.

Hard tick A tick in the family Ixodidae; known as hard because of the hard scutum on the dorsal surface.

Hemelytra A hardened, protective part of the forewing of some Hemiptera bugs.

Hemiptera The order of insects containing the triatomine bugs, the bed bugs and also many plant feeding bugs; characterized by having the mouthparts as a long piercing proboscis folded under the head.

Hide The skin of vertebrates when used for making leather, usually from cattle.

Hippoboscid A dipteran (fly) of the family Hippoboscidae, for example the louse flies and keds; these feed in close association with their hosts and have larviparous reproduction.

Holarctic Distributed in both Nearctic and Palaearctic Regions.

Holoptic In the Diptera; having eyes close together or touching above, typical of males.

Host An organism on which a parasite feeds.

Humeral pit A paired depression on the anterior and dorsal surface of the thorax of Ceratopogonidae biting midges.

Humid Having a high water content in the atmosphere, measured as per cent humidity relative to saturation with water, or as saturation deficit.

Hymenopteran A bee, wasp, hornet or ant, of the insect order Hymenoptera.

Hypersensitivity Immunological reactions of various types which lead to pathological conditions; reactions to materials such as saliva from arthropods often produce type 1 or type 4 hypersensitivity reactions in the skin; a type of allergy.

Hypopygium A part of the external genital apparatus of male tsetse (*Glossina*).

Hypopleuron A plate on the lateral surface of the thorax of Diptera (flies), the presence or absence of a row of bristles on this is important for identification; also known as the meron.

Hypostome A component of the mouthparts of mites and ticks, it is large and serrated in ticks and it forms part of the blood sucking tube situated between the palps.

Idiosoma The main part of the body of mites and ticks, it has the legs, eyes, brain, gut and other internal organs.

Insect Insects form a class (Insecta) of the pyhlum Arthropoda; the insects have a body as head, thorax and abdomen, with three pairs of legs on the thorax and often with wings.

Instar A stage in the life cycle of an anthropod, when it has hatched from the egg; usually larvae, nymph or pupa, or adult.

Integument The outer protective layer of mites, ticks and insects; consists of the outer dead layers of cuticle and the inner living layer of epidermis.

Ischnoceran Lice in the sub-order Ischnocera, chewing lice such as *Damalinia* on mammals and *Goniodes* on birds.

Ixodid A tick in the family Ixodidae, known as the hard ticks because they have a hardened plate, the scutum, on the dorsal surface.

Labella A paired organ (*singular* labellum) in the form of large lobes at the end of the labium of the mouthparts of Diptera (flies); well developed in muscid and tabanid flies where they are used like a sponge for feeding, or small and with prestomal teeth as in *Stomoxys* where they are piercing organs.

Labium The central part of the mouthparts of Diptera (flies) and other insects; it forms either a supporting sheath around the piercing mouthparts as in mosquitoes or is part of the piercing mouthparts as in tsetse.

Labrum An organ of the mouthparts of dipteran flies which forms part of the tube for sucking food.

Larva The first in the series of active immature stages of the life cycle of arthropods; they hatch from the egg and after growth develop into nymphs or pupae.

Larviparous Having reproduction in which an egg develops in the female to form a larva which then develops fully within the female, the larva is then deposited on the ground where it burrows and forms a puparium inside which it develops into a pupa; this happens in tsetse and hippoboscid flies.

Lateral On the body surface of an animal that forms the sides between the dorsal and ventral surfaces.

Lateral suture A line or ridge in the integument of *Argas* ticks which gives the appearance of dividing the dorsal surface from the ventral surface.

Leg The walking limb of arthropods, consisting of segments with a sclerotized integument forming the exoskeleton joined by flexible areas of integument; typically consists of the coxa at the base, the small trochanter, the large femur, the tibia and the outer tarsus.

Macrotrichiae Long hairs on the wings of insects, particularly on ceratopogonid midges.

Mamillae Small protrusions or bumps in the integument of *Argas* and *Ornithodoros* ticks.

Mandible A paired organ of the mouthparts of insects, may be a needle like piercing organ as in mosquitoes or a tooth-like biting organ as in chewing lice.

Mange Skin disease caused by infestation with mites.

Mastitis Inflammation of the udder, usually due to bacterial infection.

Maxilla A paired organ of the mouthparts of insects, often developed for piercing or cutting.

Medium This word is used in microscopy for the transparent and adhesive material that is used to mount a specimen on a microscope slide.

Meral rod A vertical thickening of the integument of the mesopleuron of fleas; also known as the pleural rod.

Mesopleuron A plate on the lateral surface of the thorax of insects, in cyclorrhaphan flies it has a characteristic row of bristles, in some fleas it has a meral rod.

Metamorphosis The change from the first immature stage of the life cycle to the adult stage; in the Exopterygota the change is gradual through the stages of larvae and nymphs and metamorphosis is called incomplete; in the Endopterygota the change is concentrated in the pupal stage and is called complete because the larva is changed totally into the adult.

Metastigmata A sub-order of the order Acarina; ticks, usually known as the sub-order Ixodida.

Microclimate The conditions of atmospheric temperature and humidity found on a small scale in the resting or developmental sites of living organisms.

Micropredator (*see* **Predator**).

Microtrichiae the very small extensions of the cuticle on the wings of some insects, like small setae, for example in Ceratopogonidae or non-biting midges.

Midge An informal term for a ceratopogonid, usually a biting midge as distinct from gnats and similar small flies of the family Chironomidae, or non-biting midges which are not blood feeders.

Muscid A fly of the family Muscidae, typically the house fly and similar *Musca* species.

Muscine A fly of the sub-family Muscinae, here in contrast to the sub-family Stomoxyinae

Myiasis The infestation of flesh and other tissues of living vertebrate animals by the larvae of Diptera (flies), typically calliphorids and oestrids.

Nagana Infection of domestic animals with trypanosome protozoa species transmitted by tsetse.

Nausea The sensation associated with being sick.

Nearctic Region The area of distribution of animals comprising Greenland, Canada, USA including Alaska, northern Mexico; it is separated from the Palaearctic Region by the Bering Strait.

Neotropical Region The area of distribution of animals comprising southern Mexico, central and south America.

New World The Americas, the western hemisphere of the world.

Nit Egg of louse cemented on hair of host.

Nymph The second or later in a series of immature stages of mites, ticks and exopterygote insects; between larva and adult and may consist of several nymphal stages.

Objective lens The lens on a compound microscope closest to the specimen, it makes the primary magnification which is then further magnified by the eyepiece which is the lens closest to the observer.

Obligate Specialized to live in very specific conditions; an obligate parasite is able to live only as a parasite and will not survive without its live host.

Ocelli (*singular* ocellus) Simple eyes, typically in a set of three on the head of insects between the much larger and more complex compound eyes.

Ocular Concerning or associated with eyes.

Oestrid A dipteran (fly) of the family Oestridae, a group specialized for myiasis, for example *Oestrus ovis*.

Old World Europe, Asia, Africa, the eastern hemisphere of the world.

Ophthalmomyiasis Myiasis affecting the eye.

Order A group in the classification of animals (for example Diptera) between class (for example Insecta) and family (for example Muscidae).

Oriental Region The area of distribution of animals comprising Pakistan, India, Bangladesh, south east Asia, southern China, Malaysia, Philippines and western Indonesia; the separation from the Australasian Region is the Strait of Makassar between Borneo and Sulawesi using the system of Wallace, the system of Weber includes Sulawesi in the Australasian Region.

Otitis Any inflammatory condition of the ear.

Ovigerous female A mite which is producing eggs, typically in the astigmatid mites such as *Psoroptes*.

Palaearctic Region The area of distribution of animals comprising Iceland, Europe, north Africa including the Sahara desert, Russia, countries of central Asia, countries of the Middle East, central and northern China, Japan, Korea; it is separated from the Nearctic Region at the Bering Strait.

Palp Paired organ associated with the mouthparts of arthropods, like a small limb comprising several segments and having sensory or handling functions.

Pantropical Distributed in both New World and Old World tropics.

Paragenital sinus The external genital opening of female bed-bugs.

Paralysis An inability to move, usually due to incorrect function of the nervous system.

Paramere The external genital organ of male bed-bugs.

Parasite An organism that lives on or in another organism called the host, the relationship harms the host; if the parasite remains on the surface it is an **ectoparasite**, if it invades internal tissues it is an **endoparasite**; blood sucking arthropods that only remain on the host during feeding are **micropredators**.

Parasitic Being specialized for life on or in another organism at the expense of that organism.

Paratergal Hardened plates of the integument on the lateral surface of the abdomen of some insects, typically on lice.

Parma A central bulge in the outline of the posterior margin of ticks, the central festoon; in engorged males it may expand to form the caudal appendage.

Pathogenic Causing disease.

Pedicel The stalk of the sucker which is an adhesive organ at the end of the legs of mites.

Pediculosis Heavy infestation of humans with body lice.

Perianal The area around the anus.

Perineum The area between the anus and the genital opening.

Peritreme A paired organ associated with the stigma or air breathing pore of mites of the Mesostigmata.

Phlebotomine A dipteran (fly) of the sub-family Phlebotominae in the family Psychodidae, small biting flies often known as sandflies but not the same as biting midges.

Phoresy The movement of an animal between places by clinging to another animal, as in the movement of eggs of *Dermatobia* flies attached to the bodies of mosquitoes.

Pleural rod A vertical thickening in the mesopleuron of fleas; also known as the meral rod.

Posterior The portion of the body of an animal at the end furthest from the head.

Postnotum An area of the thorax of dipteran flies, at the posterior and dorsal area next to the abdomen.

Postscutellum An area of the thorax of diptera (flies), underneath the scutellum which is the projecting posterior and dorsal margin of the thorax.

Predator An animal which obtains food by direct attack upon the live body of other animals which are known as the prey; usually involves killing the prey; the term micropredator is used for blood feeding arthropods that feed from their prey but do not stay on them like ectoparasites or kill the prey in order to feed.

Prespiracular Anterior to the spiracle, bristles of diagnostic importance in mosquitoes are in this position.

Prestomal teeth Structures at the end of the labella of some Diptera (flies) which scrape at food surfaces.

Pretarsus The last segment of the leg of mites, often formed into an adhesive sucker.

Proboscis Mouthparts which are elongated into a piercing and sucking tube.

Pronotum An area of the thorax of Diptera (flies), at the anterior and dorsal margin, near the head.

Propleuron An area of the thorax of Diptera (flies), at the anterior and central to ventral margin, near the coxa of the foreleg.

Protozoa A group of microorganisms some of which are parasitic, with unicellular structure and a membrane bound nucleus, often larger than bacteria and often have sexual reproduction, for example *Trypanosoma brucei* transmitted by tsetse and causing trypanosomosis.

Pruritus Itching, irritation of the skin.

Pseudopod A false leg, a protuberance on a larva used in locomotion.

Psoroptic Referring to mites of the family Psoroptidae.

Pterygota A sub-class of the class Insecta; all those insects which either have wings or are thought to have evolved as wingless insects from species which were winged.

Pubescent Recently become capable of reproduction, the pubescent females of some mites such as *Psoroptes* have a characteristic appearance with copulatory protuberances; another related meaning is to be covered in short soft hairs.

Pulvillus The adhesive pad between the claws at the end of the legs of ixodid ticks.

Pupa The stage between larva and adult in insects of the Endopterygota, it is this stage that has a complete metamorphosis.

Puparium The hard case that is formed (by pupariation) from the integument of the third larval stage of higher Diptera (flies); inside this case the larvae changes to a pupa; the puparium is often incorrectly called the pupa.

Pupation The process of changing from a larva into a pupa, including forming a puparium.

Pupipara A group of families of flies such as *Hippobosca* which have fully grown larva produced by the females.

Pustular Forming spots or pustules in the skin like pimples, as in some

198 GLOSSARY

forms of demodecosis.

Pygidium An area of sensory setae at the posterior of the abdomen of fleas; also known as the sensilium.

Reservoir An animal infected with a disease pathogen in such a way that infection can be passed to vectors and other hosts, while not necessarily suffering clinical disease.

Rhinitis An inflammation of the internal surfaces of the nose.

Rickettsia A group of parasitic microorganisms intermediate in size between bacteria and viruses, without a cellular structure, commonly transmitted by arthropods, for example *Rickettsia typhi* transmitted by *Xenopsylla* rat fleas and causing typhus in humans.

Rostrum The long piercing mouthparts of hemipteran bugs, also known as the proboscis.

Saliva The saliva of blood sucking arthropods is produced in paired salivary glands and is pumped into the blood feeding wound, it contains a complex mixture of agents to assist in feeding and in ticks the salivary glands also produce a cement to fix to the host's skin; microorganisms that are transmitted by arthropods often develop in the salivary glands and pass out during salivation.

Sarcoptic Referring to mites of the family Sarcoptidae.

Scab Skin disease caused by infestation with psoroptic mites.

Scale A moveable extension of the cuticle of arthropods which has a flat shape, a type of seta; butterflies and moths and the wings of mosquitoes have numerous scales giving them a clothed and coloured appearance; also flat spines may be called scales.

Sclerotized Hardened; the integument of mites, ticks and insects may be soft for flexibility at the joints or hardened into the exoskeleton or protective plates.

Screwworm Informal name for the larvae of *Cochliomyia* and *Chrysomya* dipteran flies which cause myiasis; these larvae have rings of spines and a tapered body which makes them look like carpentry screws.

Scutellum An area of the thorax of diptera (flies), at the posterior dorsal margin between the wings, often a large bulge with long bristles.

Scutum The sclerotized plate on the dorsal surface of ixodid ticks and some mites, often known as the dorsal shield, it is large in males, smaller and anterior in position in females; argasid ticks do not have a scutum.

Sebaceous gland A gland in the skin of mammals associated with the hair follicle, produces sebum.

Sebum An oily secretion from the sebaceous glands which spreads on the skin and hair, with many protective functions.

Sensilium An area of sensory setae on the posterior abdomen of fleas; alternatively the pygidium.

Seta A long thin structure projecting from the integument of mites, ticks and insects; produced from an epidermal cell and flexible at the base;

often called hair because of the similar appearance and function to mammalian hairs; large setae are called bristles.

Setose Having many setae, hairy in appearance.

Simuliid A dipteran (fly) of the family Simuliidae, typically blackflies of the genus *Simulium*.

Soft tick A tick of the family Argasidae, known as soft ticks because they do not have a hard scutum but they are very tough ticks and often able to survive harsher conditions than hard ticks.

Species A population of living organisms that freely interbreed under natural conditions; usually the smallest unit of classification of living organisms; the species name is given with the genus (for example *Glossina*) and the species (for example *morsitans*) and often abbreviated as *G. morsitans*; if the classification is complex there may be sub-species names, for example *G. morsitans morsitans, G. morsitans submorsitans, G. morsitans orientalis*; it is common to give the name of person (author of the name, or authority) who first described the species, for example *G. m. morsitans* Westwood; there are often great practical difficulties in defining species, which is the special task of taxonomists.

Spine In arthropods this means a small sharp projection from the cuticle but without the flexible base that a seta has.

Spiracle The plate on the integument of insects and ticks where the internal air breathing tubes (tracheae) come to the surface.

Spirochaetosis Infection with spirochaete bacteria which have a spiral shape, for example *Borrelia anserina* in birds causing avian spirochaetosis.

Spur A long sharp projection of the integument from the leg of an insect or tick in a characteristic position; distinct from a seta or bristle because it is formed from many epidermal cells.

Squama A lobe of the wing of many Diptera (flies), usually as two squamae on each wing, between the alula and the thorax.

Squamous A condition of the skin like thick scales, as in some forms of demodecosis.

Sternopleuron An area of the thorax of Diptera (flies), in the anterior dorsal area.

Stigmata (*singular* stigma) The openings on the integument of some mites where the internal air breathing tubes (tracheae) come to the surface; has same function as spiracle.

Stomoxyine Flies of the sub-family Stomoxyinae, here in contrast to muscine flies.

Stress A harmful effect on the physical or mental condition of an animal; arthropods cause stress by irritation, damage, toxic or allergic effects.

Striation A fine groove on the surface of the integument of some mites and ticks; occurs as many striations making a complex pattern like a fingerprint.

Strike Infestation with larvae of blowflies; superficial myiasis.

Stylostome A feeding tube produced at the mouthparts of trombiculid mites, it penetrates the skin of the host.

Subcutaneous Beneath the skin of vertebrates.

Suck In this context to suck means to take in blood and other fluids from a host by a pumping action of mouthparts in the shape of a tube; it is a more accurate description of blood feeding by arthropods than the term bite which refers to cutting up solid food.

Sucker Name commonly given to the adhesive structure at the end of legs such as the pulvillus or caruncle.

Surra Infection of domestic animals with trypanosome protozoa of species not transmitted by tsetse, typically *Trypanosoma evansi* infection.

Sweating sickness A condition of the skin of cattle and other hosts of ticks caused by toxic components of the saliva of the feeding female ticks; known as moist eczema or exudative dermatitis.

Tabanid A dipteran (fly) of the family Tabanidae, typically the horse flies, deer flies and clegs.

Tarsus The outermost section of the legs of arthropods; may be a single segment as in mites and ticks or in subsegments (usually five) in insects.

Taxonomic Concerning the identification, naming and grouping of living organisms; classification.

Thorax The central body section of insects, bearing three pairs of legs and one or two pairs of wings.

Tibia An outer segment of the legs of arthropods, between the femur and tarsus which is at the end of the leg.

Toxaemia Illness caused by poisoning.

Toxic Poisonous.

Trachea Air breathing tube in the body of an arthropod.

Transmit When a pathogenic organism passes from one host to another host through or on an arthropod the arthropod is said to transmit the organism; such an arthropod is a vector; if the pathogen has essential development within the vector the transmission is called biological, if the pathogen is simply transferred on the mouthparts the transmission is called mechanical.

Triatomine A blood sucking bug in the sub-family Triatominae of the family Reduvidae, of the insect order Hemiptera.

Trochanter Leg segment situated between the coxa and the femur (may be fused with the femur).

Trombiculid A mite in the family Trombiculidae of the sub-order Prostigmata, characterized by having the larva blood feeding and the adults free-living.

Trypanosomiasis Disease caused by infection with trypansome protozoa; sometimes spelt trypanosomosis.

Tubercle A large rounded projection from the surface of an arthropod.

Udder The milk producing organ of cattle, sheep, goats etc.

Urticaria Inflammation and irritation of the skin associated with allergic or similar reactions.
Vector An arthropod that transmits a pathogenic organism from one host in which disease may occur to another host.
Vein Tubes of cuticle in a network which give support to the wings of insects.
Venomous Poisonous for the purpose of controlling prey or for defence.
Ventral In arthropods the ventral surface is the side of the body where the legs project and make contact with the surface; in vertebrates it is the side of the body away from the spine.
Ventral plate Hardened areas of the integument on the ventral surface of the idiosoma of male *Ixodes* ticks and of some insects such as triatomine bugs; the term may also be used for the anal plates of males of other genera of ticks.
Vertebrate An animal with a spinal cord of nerves and supported by a spine of vertebral bones; includes birds and mammals.
Warble Swelling in skin caused by infection with larvae of flies causing furuncular myiasis, typically infestation with larvae of *Hypoderma*.
Zoogeographical region The surface of Earth can be divided into areas where the groups of animals found are characteristic and can be related to separate evolutionary development in areas isolated from each other, this is convenient for grouping animals by geographic area.

BIBLIOGRAPHY

Note: These are arranged by their main subject. Many of them are no longer in print.

IDENTIFICATION

Burgess, N.H.R and Cowan, G.L.O (1992) *A Colour Atlas of Entomology*, Chapman & Hall, London.
Furman, D.P. and Catts, E.P. (1982) *Manual of Medical Entomology*, Cambridge University Press, UK.
Georgi, J.R. (1985) *Parasitology for Veterinarians*, W.B.Saunders Co., USA.
Hoogstraal, H. (1956) *African Ixodoidea* Vol. 1, *Ticks of the Sudan*, US Naval Medical Research Unit No.3, Cairo, Egypt. (vol 1 only)
Howell, C.J., Walker, J.B. and Nevill, E.M. (1978) *Ticks, Mites and Insects Infesting Domestic Animals in South Africa* (Part 1 Descriptions and Biology), *Science Bulletin* No. 393, Department of Agricultural Technical Services, Pretoria, Republic of South Africa.
Ledger, J.A. (1980) *The Arthropod Parasites of Vertebrates in Africa South of the Sahara* Vol 4, *Phthiraptera* (Insecta), South African Institute for Medical Research, Johannesburg, Republic of South Africa.
Macy, R.W. and Berntzen, A.K. (1971) *Laboratory Guide to Parasitology*, Charles C. Thomas Publisher, Illinois, USA.
Matthyse, J.G. and Colbo, M.H. (1987) *The Ixodid Ticks of Uganda*, Entomological Society of America, Maryland, USA.
McDaniel, B. (1979) *How to Know the Mites and Ticks*, W.C. Brown Co., USA.
Peters, W (1992) *A Colour Atlas of Arthropods in Clinical Medicine*, Wolfe Publishing Ltd, UK.
Pittaway, A.R. (1991) *Arthropods of Medical and Veterinary Importance: A Checklist of Preferred Names and Allied Terms*, CAB International, UK.
Sloss, M.W. (1970) *Veterinary Clinical Parasitology*, Iowa State University Press, USA.

Smith, K.G.V. (Ed.) (1973) *Insects and Other Arthropods of Medical Importance*, British Museum (Natural History), UK.
Smith, K.G.V. (1986) *A Manual of Forensic Entomology*, British Museum (Natural History) UK and Cornell University Press, USA.
Urquhart, G.M., Armour, J., Duncan, J.L. et al. (1987) *Veterinary Parasitology*, Longman Group Ltd, London.
Whitlock, J.H. (1960) *Diagnosis of Veterinary Parasitisms*, Lea & Febiger, Philadelphia, USA.
Zumpt, F. (Ed.) (1961, 1966) *The Arthropod Parasites of Vertebrates in Africa South of the Sahara* Vol 1, *Chelicerata;* Vol 3, *Insecta* (excluding Phthiraptera), Vol 4, Phthiraptera South African Institute for Medical Research, Johannesburg, Republic of South Africa (Vol 2 not published).
Zumpt, F. (1965) *Myiasis in Man and Animals in the Old World*, Butterworths, London.

COLLECTING

Bram, R.A. (Ed.) (1978) *Surveillance and Collection of Arthropods of Veterinary Importance,* United States Department of Agriculture, Handbook No. 518, USA.
Cogan, B.H. and Smith, K.G.V. (1974) *Instructions for Collectors* No 4a, *Insects*, British Museum (Natural History), London.
Oldroyd, H. (1973) *Collecting, Preserving and Studying Insects*, Hutchinson & Co. Ltd, UK.
Service, M.W. (1993) *Mosquito Ecology: Field Sampling Methods*, Elsevier Applied Science, UK and USA.
Southwood, T.R.E. (1978) *Ecological Methods: With Particular Reference to the Study of Insect Populations*, Chapman & Hall, London.
Upton, M.S. (1991) *Methods for Collecting, Preserving and Studying Insects and Allied Forms*, CSIRO Division of Entomology, Australia.

BIOLOGY AND CONTROL

Alexander, J.O'D (1984) *Arthropods and Human Skin*, Springer-Verlag, Germany.
Anon. (1989) *Geographical Distribution of Arthropod Borne Diseases and Their Vectors*, World Health Organization, Geneva, Switzerland.
Anon. (undated) *Ticks and Tick Borne Disease Control*, Vol 1, Food and Agriculture Organization, Rome, Italy.
Anon. (undated) *Training Manual for Tsetse Control Personnel*, Vols 1, 2 and 3, Food and Agriculture Organization, Rome, Italy.
Burgess, N.R. (Ed.) (1981) *John Hull Grundy's Arthropods of Medical*

Importance, Noble Books Ltd, UK.
Busvine, J.R. (1980) *Insects and Hygiene*, Methuen & Co. Ltd, UK.
Curtis, C.F. (Ed.) (1991) *Control of Disease Vectors in the Community*, Wolfe Publishing Ltd, UK.
Gordon, R.M. and Lavoipierre, M.M.J. (1962) *Entomology for Students of Medicine*, Blackwell Scientific Publications, Oxford.
Goreham, J.R. (1987) *Insects and Mite Pests in Food: An illustrated key*, Vols 1 and 2, USDA and US Department of Health and Human Services, Agriculture Handbook 655, Washington.
Greenburg, B. Flies and Disease, Vol 1 (1971), *Ecology, classification and biotic associations;* Vol 2 (1973), *Biology and disease transmission*; Princeton University Press, Princeton, New Jersey, USA.
Harwood, R.F. and James, M.T. (1979) *Entomology in Human and Animal Health*, Macmillan Publishing Co., USA.
Kettle, D.S. (1990) *Medical and Veterinary Entomology*, C.A.B., UK.
Lancaster, J.L, and Meisch, M.V. (1986) *Arthropods in Livestock and Poultry Production*, Ellis Horwood Ltd, UK.
Lane, R.P. and Crosskey, R.W. (Eds) (1993) *Medical Insects and Arachnids*, Chapman & Hall, London.
Lapage, G. (1968) *Veterinary Parasitology*, Oliver & Boyd, Edinburgh.
Patton, W.S. and Evans, A.M. (1929) *Insects, Ticks, Mites and Venomous Animals of Medical and Veterinary Importance*, Part 1: Medical, Grubb Ltd, UK.
Patton, W.S. (1931) *Insects, Ticks, Mites and Venomous Animals of Medical and Veterinary Importance*, Part 2: Public Health, Grubb Ltd, UK.
Schofield, C.J. (1994) *Triatominae: Biology and control*, Eurocommunications Publications, Bognor Regis.
Service, M.W. (1980) *A Guide to Medical Entomology*, Macmillan Press Ltd, UK.
Service, M.W. (1986) *Lecture Notes on Medical Entomology*, Blackwell, Oxford.
Sewell, M.M.H. and Brocklesby, D.W. (Eds) (1990) *Handbook on Animal Diseases in the Tropics*, Baillière Tindall, London, UK.
Soulsby, E.J.L. (1982) *Helminths, Arthropods and Protozoa of Domestic Animals*, Baillière Tindall, London.
Williams, R.W., Hall, R.D., Broce, A.B. and Scholl, P.J. (Eds) (1985) *Livestock Entomology*, John Wiley & Sons, USA.
Youdeowi, A. and Service, M.W. (Eds) (1983) *Pest and Vector Management in the Tropics*, Longman Group Ltd, UK.

INDEX

Page numbers in **bold** type refer to illustrations

Abscess 115, 118
Acanthocheilonema, see Dipetalonema
Acaridida, *see* Astigmata
Acarina xix, 1
Acarus **13**
Actinedina, *see* Prostigmata
Aedes 70, 74, **75**
Aegyptianella pullorum 31, 183
African horse sickness 64
African swine fever 32
Age-grading 176
Allergy 3, 13, 14, 19, 40, 67, 181
Allodermanyssus 21
Alopecia 6, 16, 181
Alphitobius 167
Amblycera xx, 132, 143
Amblyomma **37**
Analgidae xix, **14**
Anaemia 84, 149, 181
Anaplasma
  centrale 36
  marginale 36, 44, 84, 183
Anaphylaxis 67, 166, 181
Androctonus 163
Androlaelaps 23
Animal bait trap 172
Anocentor **43**
Anopheles **57**, 68, 70, **71**, 73
Anophelinae 68, 73
Anoplura xx, 132
Ant **166**
Antenna 1, 56, 131, 143, 147

Anthrax 84, 183
Antricola 48
Aphonopelma 164
Apis mellifera 166
Apodeme 3
Aponomma 48
Apterygota 51
Arachnida xix, 1, 163
Araneae, *see* Araneida
Araneida xix, **164**
Argas **30**
Argasidae xix, 1, 25
Armillifera 170
Arthropoda xix
Aspirator **172**
Asthma 13, 14, 181
Astigmata xix, 3
Atherigona 90, 93
Atherix 85
Atrax 164
Auchmeromyia 115, **117**
Austroleptis 85
Austrosimulium 65

Babesia
  bigemina 36
  bovis 36, 48
  canis 42, 44
  divergens 48
  equi 42, 44
  major 46
  microti 48

Babesiosis 36, 40, 44, 48
Bacillus anthracis 84
Baffle trap **172**
Bait trap 172
Bartonella bacilliformis, see Carrion's disease
Bee 166
Beetle **167**
Berne, *see Dermatobia*
Biconical trap **174**
Bironella 68
Blackfly 65
Blanket drag **173**
Blister 167, 181
Bloodmeal test 176
Blowfly 103
Bluetongue 64
Boil, *see* Myiasis, furuncular
Boophilus 34, **35**
Booponus 166
Borrelia
  anserina 31
  burgdorferi 48
  duttoni 32
  recurrentis 133
Bot 126, 199
Bovicola, see Damalinia
Brachycera xx, 56, **57**, 78
Brugia 70
Bug
  bed 159
  cone-nose 155
Buthacus 163
Buthus 163
Butterfly **167**

Calliphora **57**, 103, 104, **105**, 108
Calliphoridae xx, 86
Callitroga see Cochliomyia
Calypter 53
Canker 5
Capitulum 1
Carrion's disease 60
Caruncle 3
Caterpillar 167
Centipede 165
Centruroides 163
Cephalopina 119, **121**

Cephalopsis, see Cephalopina
Cephenemyia **119**
Ceratophyllidae xx, 147
Ceratophyllus 152
Ceratopogonidae xx, 60
Chaga's disease 158
Chagasia 68
Chaoboridae 68
Chelicera 1, 34
Chelopistes 131
Cheyletiella **17**
Cheyletidae xix, 17
Chilopida xx, 166
Chigger 15
Chironomidae 61, 68
Chloropidae 95
Chorioptes **5**
Chrysomya 108, **109**
Chrysops **57**, 78, **81**, 84
Cimex **159**
Cimicidae xx
Classification xix
Cleg 78
Climate xviii
Cnemidocoptes, see Knemidokoptes
Cnephia 65
Cobboldia 123
Cockle 90
Coleoptera xx, 167
Collection methods 171
Colorado tick fever 44
Comb 148
Coquilletidia 74, **75**
Cordylobia 115, **117**
Corynebacterium pyogenes 94
Cosmiomma 49
Cowdria ruminantium 38
Cowdriosis 38
Coxa 3
Coxiella burneti 40, 44
Crab 169
Crustacea xx, 169
Cryptostigmata xix, 24
Ctenidia 148
Ctenocephalides **148**
Cuclotogaster 142, **143**
Culex 55, 70, **71**, **72**, **73**, 74

INDEX   **207**

Culicidae xx, 68
Culicinae 68, 73
Culiocoides 60, **62**
Culiseta **77**
Cuterebra 126, **128**
Cuterebridae xx, 126
Cuticle 1
Cyclops **169**
Cyclorrhapha xx, 56, **57**, 85
Cytocetes phagocytophila 48
Cytodites 9
Cytoditidae xix, 9

Damalinia **138**
Deer fly 78
Delusions of infestation 170
Demodicidae xix, 16
Demodecosis 16
Demodex **16**
Dengue 70
Depluming 9, 14
Dermacentor 43
Dermanyssidae xix, 20
Dermanyssus **20**
Dermatitis 13, 16, 19, 40, 67
Dermatobia 126, **127**
Dermatophagoides **13**
Dermatophilosis 38
Diachlorus 78
Dipetalonema
  perstans 64
  streptocerca 64
  reconditum 149
Diphyllobothrium latum 169
Diplopoda xx, 165
Diptera xix, 51, 52
Dipylidium caninum 139, 149
Dirofilaria immitis 70
Dixidae 68
Dolichovespula 166
Dracunulus medinensis 169
Drosophila 168
Dysentery 93

East Coast fever 42
Echidnophaga **151**
Ehrlichia bovis 40
Ehrlichiosis 40, 42

Elaeophora schneideri 84
Electrocutor trap 174
Elephantoloemus 116
Emergence trap 175
Encephalitis
  Eastern 70
  equine 70
  Japanese 70
  Powassan 48
  Russian spring–summer 44, 46, 48
  St Louis 70
  tick borne 44, 46, 48
  Western 70
Endopterygota 51
Entomophobia 170
Epidermis 1
Ephemeral fever 64
Equine infectious anaemia 84, 184
Euschoengastia 15
Examination of specimens 178
Exopterygota 51
Eye fly **95**
Eye moth **168**

Fannia 90, 93
Felicola **140**
Fever
  Africa swine 32
  Colorado tick 44
  East Coast 42
  ephemeral 64
  haemorrhagic 40, 48
  Mexican spotted 44
  Papatasi 60
  Q- 40, 44
  relapsing 32, 133
  Rift Valley 70
  Rocky Mountain spotted 38, 44
  sandfly 60
  tick borne 48
  trench 133
  yellow 70
Filariasis 64, 67, 70, 84
Flea
  cat 148
  chicken 152
  chigger 152
  dog 148

human 150
rat 149, 152
sticktight 152
Fleshflies 112
Floor maggot 115
Fly
  black 65
  blow 103
  buffalo 101
  bot 119
  deer 78
  face 92
  flesh 112
  forest **87**
  fruit **168**
  head **94**
  horn 101
  horse 78
  house 90
  louse 87
  muscine 90
  sand **58**
  screwworm 108
  stable **99**
  stomoxyine 99
  sweat **94**, **95**
  tsetse 96
  tumbu 115
  warble 119, 122
Forensic medicine 107
Forcipomyia 60, **63**
Fowl cyst mite 11
Francisella tularensis 42, 44, 46, 48, 84

Gadding 125
Gamasida, *see* Mesostigmata
Gasterophilidae xx, 122
Gasterophilus 122, **124**
Gedoelstia 119
Genital pore 27
Glossina 53, **55**, 96, **97**
Glossinidae xx, 53
Glycyphagus 12
Gnat
  biting, *see* Culex
  buffalo, *see* Simulium
  non-biting, *see* Chironomidae
  turkey, *see* Simulium

Goniocotes 140 **141**
Goniodes **141**
Guinea worm 169, 184
Gyrostigma 123

Habronema 93, 101
Haemogogus 70, 74, **76**
Haemaphysalis **45**
Haematobia 96, 101, **102**
Haematobosca 99, 101, **102**
Haematopinus **134**
Haematopota 78, 81, **82**
Haemogamasus 23
Haemolaelaps 23
Haemoproteus 65
Haemorrhagic fever 40, 48, 70, 184
Halarachnid mites **23**
Haltere 53
Heartwater 38
Heartworm 70
Hemiptera xx, 51, 155
Hepatitis 93
Heterodoxus **145**
Hippelates 95, **96**
Hippobosca 57, 87, **88**
Hippoboscidae xx, 53, **57**, 58, 85, 86, 87
Hookworm 93
Hornet 166
Hottentotta 163
Hump sore 103
Hyalomma **39**
Hydrotaea **94**
Hybomitra 78, **82**, 84
Hymenoptera xx, 53, 166, 167
Hypoderma 122 **123**
Hypostome 20, 25

Idiosoma 1
Inflammation 5, 6, 7, 8, 9, 23, 64, 167
Integument 1
Insecta xix
Instar 53
Ischnocera xx, 132, 138
Itch 13, 17, 64
Ixodes **46**
Ixodida xix, 1, 25
Ixodidae xix, 25

Jaundice 42

Ked 87
Keratoconjunctivitis 93, 185
Knemidocoptes 9
Kyasanur forest disease 46

Laelaps **22**
Laminosioptes **10**
Larva 1, 53, 56
Lasiohelea, see *Forcipomyia*
Latrodectus 164
Leishmania
   braziliensis 60
   donovani 60
   mexicana 60
   tropica 60
Leishmaniasis 60
Leiurus 163
Lepidoptera xx, 167, 168
Lepiselaga 78, **83**
Leprosy 93
Leptocimex 160
Leptoconops 60, **63**
Leptosyllidae 148
Leptotrombidium 15
Leucosis (avian) 167
Leucocytozoon 65
Lice
   biting, see Chewing lice
   chewing 143
   human body 132
   human head 132
   human pubic 133
   sucking 132
Light trap 171
Linguatula 170
Linognathus **135**
Lipeurus **142**
Liponyssoides 21
Lipoptena 87
Loa loa 84
Loasis 84
Louping ill 48
Louse flies 87
Loxosceles 164
Lucilia 103, 104, **106**, 108
Lumpy eye 122

Lungworm 169
Lutzomyia 58
Lycosa 164
Lyme disease 48
Lyperosia, see *Haemotobia*

Malaria 70
Malaise trap 174
Mallophaga 132
Mandible 138, 143
Mange 8, 64
Manitoba trap 174
Mansonella ozzardi 64
Mansonia 70, 74, 75
Margaropus 49
Mastitis 94
Meal worm 167
Megninia **14**
Meloidae 167
Melophagus 87, 88, **89**
Menacanthus **144**
Menopon 144, **145**
Meral rod 149
Mesobuthus 163
Mesostigmata xix, 20
Metamorphosis 1, 51, 53
Metastigmata, see Ixodida
Microscopy 179
Midge
   biting, see Ceratopogonidae
   non-biting, see Chironomidae
Millepede 165
Mite
   airsac 10
   akamushi 15
   astigmatid 3
   beetle 24
   bird 20
   chicken 21
   chigger 15
   cryptostigmatid 24
   depluming 9
   feather 14
   follicle 15, 16
   food 12
   fowl 22
   fowl cyst 11
   harvest 15

fur 17
house dust 13
itch 12, 17
mange 3, 8
mesostigmatid 20
oribatid 24
prostigmatid 15
psoroptic 3
rat 23
rodent 23
sarcoptic 7
scab 3
scrub 15
Moniezia 24
Moraxella bovis 93
Morellia **95**
Mosquito 56, 68
Moth 167, 168
Musca 53, **54**, **55**, 90, **91**, 93
Muscidae xx, 86, 90
Muscomorpha, *see* Cyclorrhapha
Muscina 90, 98
Myiasis
　facultative 87, 108
　faruncular 115, 118, 129
　gastric 125, 126
　nasal 112, 122
　obligate 87
　ophthalmic 122, 126
　rectal 125
　subcutaneous 34, 42, 108, 112
　superficial 93, 108, 112, 115
　urinary 93
Myobia 19
Myobiidae 19
Myocoptes 11

Nagana 98
Nairobi sheep disease 42
Nausea 48, 67
Nematocera xix, 56, 57, 58
Nematode 67, 84, 93, 99, 103
Neocuterebra 128
Neolipoptena 87
Neoschoengastia 15
Nets (sampling) 172
Nodule 10, 16
Nosomma 49

Nosopsyllus **152**
Notoedres 8
Nuche, *see Dermatobia*
Nuisance 93, 94, 95, 96
Nycteribiidae 87
Nymph 1, 131

Oedemagena 122
Oestridae xx, 86
Oestrus 119, **120**
Odour bait 173
Onchocerca
　gibsoni 68
　gutturosa 68
　volvulus 67
Onchocerciasis 65, 67
Ophyra 90, 93, 94
Ophthalmomyiasis 122, 126
Oribatid mite **24**
Ornithodoros 31
Ornithonyssus **21**
Oroya fever, *see* Carrion's disease
Otitis 5, 7, 20, 33, 42
Otobius 1, **33**
Otodectes **6**
Oxylipeurus 131

Paedurus 167
Palp 1, 4, 26, 57, 143
Pamphobeteus 164
Pancytopenia 42
Panic 125, 182
Pangoniinae 78
Panstrongylus **157**
Papatasi fever, *see* Sand fly fever
Parafilaria 93
Paragonimus 169
Paralysis 32, 40, 42, 44, 48, 182
Pathogen isolation 176
Pediculosis 133
Pedicel 3
Pediculus 132, **133**
Pedipalpida xix, 164
Pentastomida xx, 170
Pericoma 59
Peritreme 20
Phaenicia, see Lucilia

Pharyngobolus 119
Phlebotomus **58**
Phoneutria 164
Phormia 103, **106**, 108
Phoresy 129
Phthiraptera xx, 131
Pink eye 93
Plague 149
Plasmodium 70
Platycobboldia 123
Pleural rod 149
Pneumocoptes 11
Pneumonyssoides 20, 23
Pneumonyssus 20, **23**
Powassan encephalitis 48
Pogonomyrmex 167
Poisoning 163, 164, 166
Polio 93
Polistes 166
Polyplax **137**
Preservation 176
Pretarsus 3
Prosimulium 65
Prostigmata xix, 15
Protocalliphora 103
Protophormia 104
Pruritus 13, 18, 132, 134
Przhevalskiana 119
Pseudolynchia 87
Pseudotritia **24**
Psorergates **18**
Psorobia, see *Psorergates*
Psorophora **77**
Psoroptidae xix
Psoroptes 3
Psychoda **59**
Psychodidae xx, 58
Pterygota 51
Pthirus 133, **134**
Pulex 150
Pulicidae xx, 147
Pulvillus 3, 25
Pupa 51, 53
Pupipara 85
Puparium 53
Pyaemia 48
Pyemotes **18**
Pygidium 147

Q-fever 40, 44
Queensland itch 64

Raillietia 20, 21
Rash 7
Reduviidae xx
Region (zoogeographical) xviii
Relapsing fever 32, 133, 183
Rhagionidae 85
Rhinitis 13, 14
Rhinoestrus **57**, 119, **120**
Rhipicentor 49
Rhipicephalus **41**
Rhodnius **157**
Rickettsia
  akari 21
  australis 48
  conori 40, 42, 46, 48
  mooseri 186
  prowazekii 133
  rickettsii 38, 44
  sibirica 44, 46
  tsutsugamushi 16
  typhi 150
Rickettsiosis, see *Rickettsia* and Typhus
Rift Valley fever 70
River blindness 67
Rochalimaea quintana 133

Sabethes 74, **76**
Salmonellosis 93
Sampling methods 171
Sandfly fever 60
Sandfly 58, 60
Sarcophaga 112, **113**
Sarcoptidae xix
Sarcoptes **7**
Sentinel traps 172
Scab 5, 6
Scaly face 9
Scaly leg 9
Scorpion **163**
Scorpionida xix
Scorpiones, see Scorpionida
Screwworm 108
Scrub itch 15
Scutum 15, 25
Sensilium 147

212 INDEX

Sericopelma 164
Sheep head fly 94
Shigellosis 93
Silvius 78
Simuliidae xx, 53, 65
Simulium **57, 65**
Siphonaptera xx, 51, 147
Siphunculata 132
Siphunculina 95
Sleeping sickness 98
Smell baits 173
Snipe fly 85
Solenopotes **136**
Solenopsis 167
Solpugida xix
Solpugids **165**
Spaniopsis 85
Spider **164**
Spilopsyllus 147
Spiracle 25, 132
Spirochetosis 31
Squama 53
Stable fly 99
Staphylinidae 167
Staphylococcus 48
Stasisia, see Cordylobia
Stephanofilaria stilesi 103
Sternostoma 20
Sticky trap 174
Stigmata 3, 15
Stomoxys 96, **99**
Streblidae 87
Stress 5, 34, 36, 44, 64, 67, 70, 84, 93, 101, 103, 135, 136, 170
Strike 108
Stygeromyia 99, 101
Stylostome 16
Sucker 3, 5
Surra 84, 101
Sweat fly 94, 95
Sweating sickness 40
Sweet itch 64
Symphoromyia **85**

Tabanidae xx, 53, 78
Tabanus **57**, 78, **79, 80**, 84
Tampan 30
Tapeworm 24, 139, 149, 169

Tarsus 12, 132, 143
Telmatoscopus 59
Theileria
  annulata 40
  parva 42
  mutans 38
  taurotragi 42
Theileriosis 40, 42
Thelazia 93
Threadworm 93
Thrip **169**
Thysanoptera xx, 169
Tick
  bont 37
  bont legged 39
  blue 34
  brown ear 42
  cattle 34
  dog 42, 46
  ear 33
  fowl 30
  hard 25
  kennel 42
  lone star 37
  one-host 27, **29**
  prostriate 46
  red legged 42
  soft 25
  spinose 33
  three-host 27, **28**
  two-host 27
  zebra 41
Tick borne fever 47
Tick borne encephalitis 44, 46, 48
Tipulidae 68
Tityus 163
Tongue worm 170
Tools (entomological) 179
Torsalo, *see Dermatobia*
Toxaemia (Toxicosis) 32, 40, 42, 67
Toxorhynchitinae 68
Trachoma 93
Trench fever 133
Triatoma **156**
Triatominae 51, 155
Trichodectes 138, **139**
Trombicula **15**
Trombiculidae xix, 1, 15

INDEX 213

Trypanosoma
  brucei 98
  congolensis 98
  cruzi 158
  evansi 84, 101
  gambiense 98
  melophagium 90
  rhodesiense 98
  simiae 98
  vivax 84, 98
Trypanosomiasis 84, 98
Tsetse 53, 96
Tsutsugamushi disease 16
Tuberculosis 93
Tularaemia 42, 44, 46, 48, 84
Tumbu fly 115
Tunga **152**
Tungidae xx, 147
Typhus
  louse borne or epidemic 133
  Mexican spotted fever 44
  murine 150
  Rocky Mountain spotted fever 38
  Siberian tick 44
  scrub 16
  tick 38, 40, 42, 44, 46, 48
Tyrophagus 13

Urticaria 167

Verruga peruana, *see* Carrion's disease
Vespa 166
Vespula 53, **54**, 166
Visual trap 174
Virus
  African horse sickness 64
  African swine fever 32
  avian leucosis 167
  bluetongue 64

Chikungunya 70
Colorado tick fever 44
dengue 70
Eastern encephalitis
ephemeral fever 64
equine infectious anaemia 84
hepatitis 93
Japanese encephalitis 70
Kyasanur Forest 46
louping ill 48
Nairobi sheep disease 42
Omsk haemorrhagic fever 48
O'nyong-nyong 70
Papatasi fever 60
Powassan encephalitis 48
Rift Valley fever 70
Ross River 70
Russian spring–summer
  encephalitis 44, 46, 48
sandfly fever 60
St Louis encephalitis 70
tick borne encephalitis 44, 46, 48
Venezuelan encephalitis 70
Western encephalitis 70
yellow fever 70
Warble 199
Wasp 166
Whipscorpion **164**
Wohlfahrtia 112, **114**
Wuchereria 70

Xenopsylla **149**

Yaws 93
Yellow fever 70
Yersinia pestis 150, 151

Zoogeographical region xviii